We Are All Stardust

◆

THE EXPERIMENT

BECAUSE EVERY BOOK IS A TEST OF NEW IDEAS

We Are All Stardust

Scientists Who Shaped Our World Talk about Their Work, Their Lives, and What They Still Want to Know

◆

Stefan Klein

Translated by Ross Benjamin

THE EXPERIMENT

NEW YORK

An earlier version of this book was originally published in German by
S. Fischer Verlag GmbH, Frankfurt am Main, Germany, in 2010.

The Experiment, LLC
220 East 23rd Street, Suite 301
New York, NY 10010-4674
www.theexperimentpublishing.com

The Experiment's books are available at special discounts when purchased in bulk for premiums and sales promotions as well as for fundraising or educational use. For details, contact us at info@theexperimentpublishing.com.

Many of the designations used by manufacturers and sellers to distinguish their products are claimed as trademarks. Where those designations appear in this book and The Experiment was aware of a trademark claim, the designations have been capitalized.

Library of Congress Cataloging-in-Publication Data

Klein, Stefan, 1965-
[Wir alle sind Sternenstaub. English]
We are all stardust : scientists who shaped our world talk about their work, their lives and what they still want to know / Stefan Klein ; translated by Ross Benjamin.
pages cm
Includes index.
ISBN 978-1-61519-059-1 (pbk.) -- ISBN 978-1-61519-153-6 (ebook) 1. Scientists--Interviews. 2. Science--Philosophy. I. Title.
Q141.K64613 2015
500--dc23
2013003769

ISBN 978-1-61519-059-1
Ebook ISBN 978-1-61519-153-6
Cover design by Allison Colpoys
Text design by Pauline Neuwirth, Neuwirth & Associates, Inc.

Manufactured in the United States of America
Distributed by Workman Publishing Company, Inc.
Distributed simultaneously in Canada by Thomas Allen and Son Ltd.
First printing November 2015
10 9 8 7 6 5 4 3 2 1

Contents

Introduction

On the courage to spend a lifetime searching

◆

SCIENCE DEFINES OUR LIVES AS NEVER BEFORE. And yet we don't know much about the people who change our world with their research. It can't be that they have nothing to share. Many of the scientists I met for the conversations in this book look back on astonishing life stories. They have unusual interests and think far beyond the horizons of their respective fields. In short, as people they are just as interesting as the actors, professional athletes, or politicians whose inner lives we hear about in minute detail.

Yet we still tend to picture all scientists as Einstein, the genius who is inept at life and sticks his tongue out at the world. This lingering perception is partly scientists' own doing: They try to deny their personal side. Science aims to be objective; the self is to be kept out of it. In a scientific publication, to use the word "I" is sacrilege. And because scientists, of course, crave recognition as much as anyone else, they help construct the myth that surrounds them. Scientists may be discouraged from expressing their individual personalities too strongly in their professional world, but they can at least flatter themselves that they are scholars who hover above everyday things.

But there's another, deeper reason that the thoughts and feelings of scientists are unknown to the public: Our society views science with tunnel vision. Scientific research is rightly perceived as a source of prosperity; it has brought us effective medications, computers, and thousands of other amenities. The work of scientists in their laboratories is clearly useful, even if we don't always understand it.

But in the eyes of most people, it has nothing to do with what really moves us, with the existential questions of our lives.

To think that way, however, is to overlook the fact that science is part of our culture—like our books, our music, our movies. From its beginnings, science has explored the mysteries of our existence. And particularly in recent years, scientists have gained many insights that help us see more clearly who we are, where we come from, and what it means to be human.

For this book I have met some of the women and men to whom we owe such insights. With two exceptions, my interviews with scientists—from Europe, the United States, India, and Australia— took place between 2007 and 2012. I was encountering almost all my subjects for the first time. Usually I arranged to meet with them on two consecutive days, always in a place of their choosing. Often we spoke in their offices, but we also occasionally conversed on long walks or in restaurants, museums, or their houses. The one thing I demanded in advance from the scientists was time; generally our dialogues—some conducted in German, some in English—lasted five hours.

The interviews collected here are distilled from those conversations, edited and significantly condensed to include the most interesting passages. All but one first appeared in the magazine supplement to *Die Zeit*, the German weekly newspaper, and many interviews went on to be included in a prior German edition of this book. These interviews thus took an indirect path to their English-language publication here (in many cases, after having been translated into German for *Die Zeit*, and now translated back again). To ensure nothing has been lost in the process, every interview in this book has been reviewed and approved (and sometimes lightly updated) by the interviewee. For several conversations, I decided to take this opportunity to revisit the original, voluminous transcripts. Working directly from those, I have been able to bring to light many revealing passages not included in the former newspaper versions.

All my conversation partners enjoy a worldwide reputation in their fields, and all have made a mark by placing their research in a broader context. They include a winner of the Nobel Prize in Chemistry who has made a name for himself as a poet, a cosmologist who publicly wagers on the fate of the world in the next decades, and a physiologist who at the same time investigates the origins of civilizations in the jungles of Papua New Guinea. I have tried to give representatives of a wide variety of scholarly interests a chance to speak: A geographer, a philosopher, a social scientist, an economist, and an anthropologist offer perspectives from the humanities and social sciences. Incidentally, the selection is unabashedly subjective. I've talked to people I wanted to meet—because their lives and scientific contributions struck me as extraordinary.

The objection that in this collection white men are disproportionately represented is undeniable: Only five of my conversation partners are women, and only three are not from Europe or the United States. But this assortment is a snapshot of our time. I sought scientists with far-reaching lifetime achievements and broad perspectives, which are typically attained in the second half of a career. Among researchers of that age, women as well as people from Asia, Latin America, or Africa are still rare. Today the young talent in the laboratories is fortunately more diverse, and thus the assortment of my conversation partners would look different two decades from now.

In the conversations, I pursued two simple goals: to learn who my conversation partners are and what they do. For me, those are really just two ways of phrasing a single question, for I have never believed in the myth that scientists can disregard their personal side in their work. It seems to me almost self-evident that their life stories, not to mention their cultural roots, shape their interests. But for most of those I spoke with, my approach was extremely unfamiliar. As the neuroscientist Hannah Monyer put it, "In science, the individual doesn't count." It was astonishing, in light of her comment, how many of them opened up about themselves once we had developed a level of trust. Had they felt burdened by the need to inhibit themselves?

Not all of them were completely at ease with me, however. World-renowned scholars occupying the highest academic posts, who give lectures all the time in front of their students and at conferences, suddenly lost all their eloquence when it came time to talk about themselves. Still, they did enjoy the opportunity for a little self-expression. Only they clearly felt guilty about it—as if I had inveigled them into something unseemly. The fear of revealing a weakness through an ill-considered remark was too deep-seated.

I wasn't surprised that the conversations with Nobel laureates—their reputations secure—were among the most relaxed. At the same time, especially before the meeting with the physicist Steven Weinberg, I was more than a little nervous, as this almost legendary scientist had, with his essays and books, been a significant presence for a good twenty years of my career. There are probably few physicists of my generation the whole world over who would not revere Weinberg as a supreme authority. So I shot frenetically on my rented bike across the campus of the University of Texas at Austin, passing his department several times before I finally sat across from him, sweaty and late. After we greeted each other, I confessed to feeling intimidated. I told him how much and how early in my career his works had influenced me—and was immediately embarrassed by my words, which I feared he must have heard a thousand times before. But Weinberg's eyes lit up: "That's very pleasing to hear." From that point on, the spell was broken. Rarely have I met anyone less pretentious—and rarely anyone who so candidly admitted their own mistakes, lapses, and doubts. When you have achieved everything, you have nothing left to prove.

Each of the scientists I met commanded my respect. What inspired my admiration, however, was not so much their outstanding intelligence, which is so often ascribed to eminent scientists. Certainly I was dealing with women and men of extremely keen intellect, but there were few whose mental powers struck me as beyond reach. Nobel Prize winners are not smarter than other people, the chemist Roald Hoffmann, who is one himself, suggested in our conversation. I would add: And if they nonetheless attain heights that are impossible for others, it is not because they were born with superior brains, but because they better train their gray matter. Their intelligence was never a given; it had to be developed on a path that all my conversation partners pursued. They had dedicated their lives to the goal of discovering a few puzzle pieces of the world. It was this capacity for devotion, which shone through in each conversation, that I admired and that often moved me. Devotion can bring people the highest moments of happiness, but it exacts a heavy toll. In this series of conversations, however, the only scientists to acknowledge the cost of their commitment to cutting-edge research were the women. The fact that they alone—and none of the men—addressed this subject seems hardly a coincidence.

While the media report only the news of scientific successes, very few outsiders know the vast price in failure and disappointment at which every single triumph comes. The mysteries of nature are like a labyrinth: The solution appears only after every wrong path has been tried at least once. And even someone who, without knowing it, is on the right track must toil painstakingly for years, sometimes decades, before solving a fundamental problem. It's not intelligence that is the most important trait of a scientist, but persistence—a determination bordering on obstinacy to contend with setbacks, self-doubt, even monotony. The geneticist Craig Venter acknowledged that "behind every science experiment . . . there's a lot of tedium." He described to me an experiment of his that required isolating an adrenaline receptor as "boring" and "frustrating"—and yet, to him, thrilling. "There is truly nothing more exciting," he said, "than having an idea, thinking about a way to test that idea, and getting an answer that gives you new information about the world around us." It's a remarkable attitude to have, especially since, as Venter admits, "most science crawls along."

What, then, brings people to scientific research? What keeps them doing it? Just as no scientist is born brilliant, no one is born a scientist. On the contrary, almost all my conversation partners told me how they arrived at their current field of interest and ultimately found success largely by chance: Sarah Hrdy was doing research for a debut novel she planned to write about the Maya people when she discovered anthropology and stuck with it. Ernst Fehr, today one of the most influential economists in the world, had intended to become a priest. And Raghavendra

Gadagkar might never have risen to prominence as a leading animal behavior researcher had his attention not been aroused by the ubiquitous wasps in his college dorm in India. For others, it was the encounter with a charismatic teacher that gave their life a completely new direction. In light of these biographies, the hope that careers can be planned turns out to be unfounded. It was not farsighted thinking that helped these now-accomplished scientists get ahead but their self-confidence to overcome obstacles and go their own way.

They have retained that courage to this day. But it's not the audacity to defy the gods; rather, it's a willingness to spend a lifetime searching. Thus many of the scientists I met displayed a peculiar mixture of personality traits — strong egos that at the same time allowed them to embrace uncertainty. Beyond the vanity, beyond the desire to immortalize one's own name with a groundbreaking discovery, an impulse seemed to be at work in all of them: a pleasure in being on the way and knowing well that they will never arrive.

One of the best answers to the question of what motivates scientists is also one of the oldest. It comes from Leonardo da Vinci, who as the forefather of the modern natural sciences gives this book a historical perspective with an interview composed of quotations from his writings. For Leonardo, the thirst for knowledge was a form of love of nature and thus of life: "Love of anything is the offspring of knowledge; the more certain the knowledge, the more fervent the love." What we really understand we come to value. And because by looking closely we ultimately change ourselves, Leonardo regarded literally every object as worthy of intense engagement — the current around a pebble in a stream as much as the course of the stars.

Leonardo was a pioneer on an unknown continent. He investigated particular phenomena, each individually — the relationships between them he could at best intuit. In more than five hundred years of natural research since then, scientists have learned to see many connections; they know, for example, that the laws of the current around a pebble in a river also determine the emergence of stars in the cosmos. Thus each small-scale insight points beyond itself, the way a crack in a wooden wall can open up a view of a whole landscape. Several of my conversation partners described such an experience in almost identical words as "that wonderful moment when suddenly everything clicks." Often the most unspectacular problems lead to a much bigger mystery—and sometimes they even provide the key to solving it. Big questions hiding in small ones lie at the heart of the conversations collected here.

We Are All Stardust

◆

We Are All Stardust

Cosmologist
Martin Rees
on the beginning and end of the world
◆

MARTIN REES IS THE LAST EUROPEAN court astronomer. He is
Astronomer Royal to the House of Windsor and adviser to Her
Majesty, Queen Elizabeth II, in astronomical and scientific matters. Rees, born in 1942 in York, England, studied mathematics at
Cambridge and has been professor of astronomy there since 1973.
Over his more than forty-year career, however, Rees has sat at a
telescope only in his free time; he has made a name for himself primarily as a theorist, not least with his ingenious speculations. As
if the post of Astronomer Royal weren't enough, he served for five
years as president of the Royal Society, the oldest learned society
in the world. The Queen elevated him to Baron Rees of Ludlow,
so that he could also play a political role in the House of Lords.
Rees receives his visitor at Carlton House Terrace, the home of
the Royal Society, in the immediate vicinity of Buckingham Palace.
Over his desk hangs a huge oil painting of Sir Isaac Newton, who
was once president of the society as well. There is tea.

◆

Professor Rees, what does an Astronomer Royal do?

Oh, that's just an honorary title. It dates back to 1675, when the Astronomer Royal ran the Royal Observatory of Greenwich. Today the duties of the post are so limited that you could actually hold it posthumously.

Your investigations of quasars have yielded key evidence of the big bang. Have you ever explained to the Queen what a quasar is?

We've never talked about it. If I had the opportunity, I would tell her that they're huge black holes in the centers of galaxies that suck in gas, and before the gas plunges into the black hole, it shines more intensely than anything else in the cosmos—brighter than trillions of suns.

And what does that have to do with the big bang?

We were faced with the question of why we find more quasars the farther away from the earth we look. With my colleague Dennis Sciama I was able to show that the big bang theory provides the correct answer: Most quasars were formed in the early universe, and as the universe expands, they grow increasingly far away from us. We made this discovery in 1965.

Today we know that quasars are actually among the oldest visible objects in outer space: Recently quasars were discovered that were formed at least thirteen billion years ago. What do you find so fascinating about this phenomenon?

The physics at the edge of a supermassive black hole is really interesting. We can test Einstein's theory there, and remarkable effects occur. Above and below the gas vortex, streams of matter are thrust into the universe at enormous speed. The late Arthur C. Clarke, who wrote the science fiction novel *2001: A Space Odyssey*, once asked me whether those jets could be the product of a highly developed civilization.

Galactic beacons?

Perhaps.

But if the quasars were formed shortly after the birth of the universe, a civilization living in their vicinity would have barely had time to develop.

That's a problem. But other, more recently formed supermassive black holes shoot out jets as well.

Did you always want to investigate the universe?

Not at all. I began as a mathematician, but then I realized that I didn't want to pursue mathematics for its own sake. So I looked for a field where I could apply what I had learned. I was within an inch of becoming an economist! But after a year or so, quasars, which had just been discovered, turned out to be a good choice.

I once considered becoming an astrophysicist myself. But with my twenty-five years at the time, the stars seemed to me too far away—while there are so many astonishing things to explore right under our noses.

The stars are much closer to us than you might have thought. They're governed by the same natural laws as everything on earth, only under extreme conditions. After all, the cosmos is our environment. With all human beings who have ever lived we share the same view of the stars. And ultimately we ourselves are stardust.

Like everything on earth, we consist of the vestiges of long-extinguished celestial bodies.

Exactly. All the elements were formed in the stars out of hydrogen and helium through nuclear fusion. If you're less romantically inclined, you can call human beings stellar nuclear waste.

These must be exciting times for anyone probing the mysteries of the universe. "Cosmology is undergoing a revolution," your colleague Charles Bennett has declared.

Absolutely. Thanks to new telescopes, as well as the processing power of supercomputers, we can now trace the development of the universe back at least twelve billion years. That has helped us understand how the first stars and galaxies emerged from an unstructured state, and how they evolved up to the present. What I find even more exciting is that over the past ten years or so we have been discovering one planet after another outside our solar system. I hope we will soon find an earthlike planet.

With that very goal NASA plans to launch the space probe Kepler into space next February.*

That will open up new perspectives. The current instruments detect only giant planets, roughly the size of Jupiter. True, there could be life there too, but even more interesting would be a planet with conditions resembling those on earth at the time when life began.

What are the odds?

We might well find planets like ours, but I doubt that in the next twenty years we'll have enough evidence to prove that they have biospheres. Still, I would bet on life somewhere in the Milky Way—and even more on life in other galaxies.

---

* Kepler was launched in March 2009.

Even the best telescopes can't penetrate into most regions of the cosmos, which are so far away that their light hasn't had enough time since the origin of the universe to travel to us.

It's like when you climb the mast of a ship on the high seas. You can see only to the horizon, but you have to assume that the ocean beyond it extends much farther.

According to the data provided by the two satellites COBE and WMAP in recent years, we have to assume that the universe is many thousand times larger than the small area we can see. What might be going on in those hidden regions?

Satellite measurements suggest that beyond the horizon things remain the same as in the visible cosmos for a very long time. But of course we can't be certain.

How can anyone doubt the existence of extraterrestrial life in the face of those dimensions? The probability of life sprouting up somewhere might be infinitesimal. But if you consider the countless solar systems of the cosmos, it seems almost sure to have happened repeatedly.

You're right. And in the next two decades we'll hopefully learn more through biological experiments about how life emerges from nonliving matter. Still, it's not logically out of the question that we're alone in our galaxy.

You once said that you would prefer that. Why?

Because it would deprive us of some cosmic self-confidence if there were others out there. On the other hand, a Milky Way inhabited by life would of course be much more interesting.

But is humanity mature enough to cope with the news of having cosmic company?

I don't think the experience would have to be that traumatic. The cultural impact might roughly equal that of the discovery of America.

Do you like science fiction?

I find many ideas in it really stimulating. That's why I advise my students to read first-rate science fiction rather than second-rate scientific publications.

In Stanisław Lem's novel *Solaris*, scientists discover a strange ocean on a distant planet. It turns out to be an intelligent organism, but the scientists find no way to communicate with it. The differences between humans and that life form are too great—a plausible scenario.

Perhaps there are in fact life forms and intelligences that we don't yet even recognize as such. But it's just as possible that other civilizations would intentionally send us a message composed in such a way that we could understand it. It's far more likely, however, that we would come across very simple life forms—or their remains. That alone would be a fascinating discovery. That's why I'm in favor of all conceivable efforts to search for extraterrestrial life.

That raises a fundamental question: Why is the nature of the universe such that life could emerge in it in the first place?

True. We can imagine all sorts of universes in which life would be impossible from the start. Ours, however, is very finely balanced. If it had, for example, consisted of fewer elementary particles, no complex structures could have come into existence. Or if it had contained too much matter, the whole cosmos would have soon recollapsed into itself. And if gravity had been only slightly stronger, the stars would have been much smaller and would have burnt out long before life could emerge.

This truly incredible balance, which first makes everything possible, can't be explained by contemporary cosmology.

That's one of the most important tasks for science in the twenty-first century: to find out whether we can explain those constants—or whether we just have to accept them.

Einstein, Heisenberg, and many other physicists have searched for a theory of everything. They have failed.

But most scientists still hope that that dream is fulfilled one day.

You advocate another solution. . . .

I don't advocate anything. I'm just open to all possibilities.

One possibility would be that our universe is only one of many, that it has siblings.

Yes.

The vast majority of them are inhospitable. But in some universes— ours among them—the natural constants happen to be constituted in such a way that life can exist.

It's sort of like how you don't necessarily need a tailored suit in order to look good. You only have to go to a big clothing store. If it has a wide enough selection, you'll always find a jacket off the rack that fits. Analogously, in a large collection of universes, there would be a high probability that one of them would be suited for the emergence of life.

Creation as a gigantic shopping center—I've never heard it described that way before. Of course, the question remains: Who or what designed the collection?

According to several plausible theories, the composition of our universe didn't follow a predetermined plan, but rather was established by chance during the big bang.

You mean the theory of inflation, of which there are several variants. The most common idea is that at the beginning of time the universe suddenly inflated. First it was much smaller than an atom, then suddenly approximately the size it is now.

Exactly. In the subatomic world there are random fluctuations. These are known as quantum fluctuations, and when the whole universe inflated, those fluctuations too expanded to a cosmic scale. . . .

And formed, so to speak, the basis for all later structures in the universe. When I talked to colleagues around ten years ago about

inflation theory, I had the impression that they regarded the idea as an interesting speculation but didn't take it all that seriously. What has changed since then?

The two satellites you mentioned provided new data on cosmic background radiation, which originated about 380,000 years after the big bang and has filled the whole universe ever since. It's a sort of afterglow of the big bang and contains the oldest information available to us from the cosmos. The new measurements have confirmed several predictions of inflation theory.

And according to some versions of inflation theory there wasn't just one big bang, but many. And with each of them a new universe was born. All these universes exist side by side. Ours would then be only part of a whole array of universes—the multiverse.

Yes. The natural laws that make life possible would then apply only to our small province.

The way so-called chaotic inflation theory describes the birth of a universe is amazing: Empty space is filled with an energy that can, under certain conditions, set off inflation. Then a cosmos inflates as if a world had arisen out of nothingness. Einstein's equations yield this result. Still, I find the idea disconcerting.

Instead of "out of nothingness," we should say "out of a vacuum," because empty space is not nothing. Inside of black holes, too, space could open up and form a new universe.

This new universe would no longer have any connection at all to ours. . . .

No.

How, then, can we tell that it even exists?

We will never be able to see other universes directly. But chaotic inflation theory makes other predictions as well—for example, about the nature of gravity in our universe—that can most likely be tested. If all of them turn out to be true, it can be assumed that the whole theory is correct. After all, you can't see black holes either, but we have enough evidence to be certain they exist.

A new universe at the moment of its birth even has its own beginning of time. That's where my powers of imagination reach their limit. How do you conceive of such theories?

I try to draw pictures or even just visualize them—to the best of my abilities. It's easier for me to deal with images than with mathematical formulas. So here we would have one universe, there we would have another, and each one looks like a small black hole of its sibling.

I'm afraid that doesn't really help me. Is it possible that in the investigation of the cosmos human beings must contend with the fact that our brains aren't really equipped to grasp this reality?

Certainly. We all have trouble imagining space in four dimensions, as the theory of relativity demands. And more recent cosmological models, like string theory, describe the universe in ten or even more dimensions! Our brains are equipped for survival in the African savanna, not for thinking about cosmology. But don't forget: We have the same problem in highly abstract quantum physics—and yet we're amazingly successful. All our electronics are based on physicists' ability to calculate quantum phenomena.

Apparently there are two meanings of "to understand." In a certain sense we understand quantum physics—that is, we can apply its laws—but we're like drivers who can steer cars but have no idea what's happening under the hood. No one knows what it actually means that, for example, an electron is a "superposition of different states": In that sense we don't understand quantum physics.

You're right. But it's still amazing that we have achieved so much in the first place in fields like quantum mechanics or relativity theory, which are so remote from our everyday intuition. And there might indeed be a line our minds simply can't cross, despite all cultural progress. True, Newtonian physics was regarded as an extremely complex field during Newton's lifetime, whereas today any reasonably bright high school student can comprehend it. But I highly doubt that it will ever be possible to teach schoolchildren relativity theory and quantum mechanics.

> Maybe Einstein was right that there is a cosmic plan. But it might also be true that human brains just aren't suited to grasp it.

It wouldn't surprise me if that were the case. On the other hand, it might turn out that the conditions in our universe are purely the result of chance, because it's only one of many. But as you said: That could raise the question of where the laws of the larger unity, the multiverse, come from. The problem wouldn't be solved, only deferred.

> That's exactly my impression of the meteoric development of cosmology over the past ten years. Whenever a question is answered, five new ones arise, which are even more unfathomable.

That's simply the nature of scientific research. The more the territory of our knowledge expands, the longer the boundary between that territory and what we don't know becomes. Many questions we are grappling with today we couldn't even have asked ten years ago. Just think of dark energy. . . .

> A mysterious force that pushes the universe apart and has nothing to do with the known forms of energy.

It was first discovered in 1998. As you say, we don't have the slightest idea about its nature. We know only that it permeates everything and that it's very powerful. If you ask what the universe is composed of, you come up with dark energy more than anything else. It comprises almost three quarters of the cosmos. And most of the remaining quarter is dark matter, a substance that emits no radiation and is therefore invisible as well. About that we know at least that it determines the structures in outer space. But we have no idea what dark matter actually is, either. And the known forms of matter? They make up no more than 4 percent of the cosmos.

It's as if we were sitting on a chocolate cake and barely knew the icing.

Yes.

How do you imagine dark matter?

It must involve a shower of particles that constantly permeates the earth—and our bodies too, of course. Probably there are no particularly fine structures in this shadow world, because there's good reason to believe that the particles don't bind together. So dark matter would be a sort of ubiquitous gas that, in certain places, condenses to some degree, exerting gravitational pull that draws together visible matter to form a galaxy. But we don't know for sure.

If that were the case, then we would be dealing with something like parallel worlds: Between dark matter and dark energy on the one hand and the known forms of matter on the other hand, there are no interactions—except for gravity.

That seems to be the case. But don't forget that gravity at the level of individual particles is very weak and therefore negligible compared with the other forces between the components of matter. However, it has a very far reach—at a cosmic scale it plays the decisive role among all interactions.

Were you dismayed when you first learned that most of the universe consists of entirely different forms of energy and matter from those that are known to us?

No. There's no reason for everything in the universe to radiate and shine. Dark matter and dark energy just don't do that.

One of the tasks of science is to liberate us from our assumptions.

Exactly.

Is it true that you attend church regularly?

I was brought up as a member of the Church of England and simply follow the customs of my tribe. The church is part of my culture; I like the rituals and the music. If I had grown up in Iraq, I would go to a mosque.

Do you experience any conflict between that and your scientific worldview?

None at all. It seems to me that people who attack religion don't really understand it. Science and religion can coexist peacefully—although I don't think they have much to say to each other. What I would like best would be for scientists not even to use the word "God."

Still, science and religion have the same basic inspiration: our wonder as humans at being part of a larger reality.

That's true. But fundamental physics shows how hard it is for us to grasp even the simplest things in the world. That makes you quite skeptical whenever someone declares he has the key to some deeper reality.

Can you believe what is preached in church?

No. I know that we don't yet even understand the hydrogen atom—so how could I believe in dogmas? I'm a practicing Christian, but not a believing one.

You don't have a strong tendency toward belief as a scientist either; nor do you seem to have much trouble with contradictions. Among fellow scientists, you're known for often working on two conflicting theories at the same time. Others conduct trench warfare over the question of whether the natural constants are the result of chance or not, whether there's one universe or many.

I find it irrational to become attached to one theory. I prefer to let different ideas compete like horses in a race and watch which one wins.

You're an intellectual gambler.

No, I'm searching for the correct answer. And in that effort emotions are simply not very helpful.

Are there any subjects you do get emotional about?

Yes, political issues. I grew up as a fairly old-fashioned socialist. I'm very worried about increasing inequality and our failure to ensure that the Third World shares in the benefits of globalization. Just think of Africa: All the resources to eliminate the poverty there exist, but they aren't deployed.

You've given humanity an even worse prognosis. You've estimated the probability that our civilization will survive the century that has just begun at no more than 50 percent. How did you come up with that number?

It's an estimate. You can calculate properly only the probability of events that occur frequently, like lightning striking or people winning the lottery. I'm going by the known threats of the past several decades. During the Cold War, for example, the probability of a nuclear catastrophe was anything but slim. Nuclear missiles were secured terribly. We just got incredibly lucky. Today the systems are better. But within the next fifty years there could be a new confrontation between superpowers. In addition, there are all the dangers that didn't exist yet at that time, like those posed by genetically engineered microorganisms. I don't find my estimate too pessimistic at all.

So you stand by your bet that in the next twenty years a million people or more will die in an attack waged with biological weapons or an accident caused by biotechnology? You put a thousand pounds on that.

Yes, indeed. Of course, I hope I lose the bet.

How did your colleagues react to your prophecies?

Almost all of them agreed with me—to my surprise. It seems to me indisputable that the threats of the twenty-first century come primarily from human beings and are greater than those of previous eras. And the fact that there are more of us than ever before and we are endangering the bases of our existence through exploitation of natural resources, climate change, and so on isn't even the whole story. There are additional risks as a result of the world becoming more and more interconnected.

What dangers does the Internet pose?

It not only expands peoples' horizons but also is capable of strongly reinforcing prejudices. Small groups with extreme views can now find like-minded people all over the world, organize themselves, and easily access technical knowledge. And because the mass media exponentially increase the psychological impact of any confused action, a handful of people can now exercise enormous power. This presents a political challenge. There have been proposals, for example, for a completely transparent society, in which everyone would monitor everyone. Not the government but your fellow citizens would be watching everything you did.

A horrifying idea.

But people can quickly get used to it. Here in England we've resigned ourselves to video cameras observing our every step. The most important thing is that these questions are debated. We don't pay enough attention to many potentially disastrous events—and don't invest enough in preventing them.

What can scientists do to change that?

They can advise politicians as experts. And as citizens they should get directly involved in politics more often.

Do you believe that scientists make more intelligent decisions than other citizens?

They bring a special perspective to things. For example, as an astrophysicist, I'm used to thinking in terms of extremely long periods of time. For many people, the year 2050 is distant enough to seem unimaginably far away. I, however, am constantly aware that we're the result of four billion years of evolution—and that the future of the earth will last at least as long. When you always have in mind how many generations might follow us, you take a different attitude toward many questions of the present. You realize how much is at stake.

Provided we manage not to squander our future: Do you have a conception of the next four billion years?

We humans of the present are certainly not the summit of Creation. Species more intelligent than us will inhabit the earth. They might even appear quite soon. These days evolution is no longer driven by slow natural development, as Darwin described it, but by human culture. So a post-human intelligence might be made by us ourselves. And I hope that our successors have a better understanding of the world.

The Genes of the Good

Evolutionary biologist
Richard Dawkins
on egoism and selflessness

◆

ARE WE BORN SELFISH? In 1976 Richard Dawkins, at the time a completely unknown biologist at Oxford, caused a stir with the publication of his book *The Selfish Gene*.

Since his spectacular debut, Dawkins has written many works eloquently championing a Darwinian worldview. In 1995, he became professor for the public understanding of science at Oxford, a position he held until 2008. Most recently, he has emerged as a fierce critic of all religion. "Darwin's Rottweiler," he has been called. But the man who received me on the terrace of his very British house turned out to be a captivating conversation partner.

◆

Professor Dawkins, on a cruise with invited guests you wore an interesting T-shirt. It said, "Atheists for Jesus." Did you discover your love for Christianity under the influence of the tropical sun?

The point I wanted to make was that Jesus was a good man, and that a man of his time had to be religious because everybody was. But I suspect that if he had the knowledge we have today, he probably would have been an atheist, and he probably would have been a good man.

He replaced the rule "an eye for an eye, a tooth for a tooth" with the commandment to love your enemies and turn the other cheek. You've written before that, in a way, his teaching is anti-evolutionary.

That's right. Anti-Darwinian, I'd say, because Darwinism is going to explain only a certain limited amount of niceness. But humans do seem to be super-nice—at least some of them, or even quite a lot of them. Maybe Jesus was one of those. And that's something we should contemplate, that's something we should try to identify, where this super-niceness comes from, and try to spread it around.

You've very eloquently made this point before, that we should culti-
vate super-niceness. But you've also written that, from a Darwinian
point of view, super-niceness would be plain dumb.

Yes, that's right, from a Darwinian point of view it would be dumb,
because it's counter-Darwinism. We have lots of examples of this.
Contraception is anti-Darwinian, adoption is anti-Darwinian, but
so is super-niceness, so is being kind, generous, giving to charity,
giving to famine relief—all those sorts of things are very un-
Darwinian and should be cultivated.

Because in your view, they wouldn't help you, nor would they help you
pass on your genes?

Yes.

Evolution is a cruel business, isn't it?

Very cruel. Darwin himself noticed that. Many people have seen
this as well.

And we have good reason to assume that Darwin's laws govern our
existence to this day.

Yes, which makes it all the more surprising that we do seem to be
capable of super-niceness.

When you wrote *The Selfish Gene* over thirty years ago, you used
strong metaphors to describe those laws of evolution. You referred to
us humans and other animals, for example, as "robot vehicles blindly
programmed to preserve the selfish molecules known as genes."
Would you still phrase this the same way?

Yes, I would.

One of your foreign publishers supposedly didn't sleep for three
nights when he first came into contact with your books, because the
message seemed to him so cold and dark. How did you actually feel
while writing them?

Not dark—exhilarated, really, by the joy of understanding. But I
never thought that you should import moral value judgment lessons from nature anyway.

But can a view of evolution as you put forth in *The Selfish Gene* really account for the selflessness that human beings actually show? Even the young Charles Darwin, sailing around the world on the HMS *Beagle*, marvelled at the naked inhabitants of Tierra del Fuego. Those "savages," as he called them, had never before encountered people from another culture. And yet Darwin praised their decency and sense of fairness. He referred to a "social instinct," though he admitted that he couldn't explain it.

I think the Darwinian—the evolutionary psychological—explanation would be to say that natural selection built into us a sophisticated calculator in the head—a kind of psychological money. Barter trade with other groups would have been an advantageous thing to do from a Darwinian point of view, and from the idea of trade, the idea of exchange, the idea of debt, the idea of a very highly developed sense of who owes what to whom, we always have a very highly developed sense of obligation, debt, gratitude, guilt—all stemming from the need for keeping a constant tally: "Have I given back to you enough to repay you for what you've given to me?" You just kept in your head how much so-and-so owes you, how much you owe him.

Robert Wright, an American writer on evolutionary psychology, even attempts to explain compassion with business sense. "The more desperate the plight of the beneficiary, the larger the IOU," he writes in *The Moral Animal*. "Exquisitely sensitive sympathy is just highly nuanced investment advice." He gives the impression that great acts of sympathy are calculated, which I find cynical and untrue.

Oh, yes, but it doesn't have to be cognitively calculated. I mean, you can look at a plant and you can say there's a very complicated calculation: How much of the available resources do you put into the roots, how much into the leaves, how much into the flowers? You could do a calculation as a biologist and say that this is optimized, that you've put just the right amount into the flowers for reproduction and the right amount into the leaves for gathering new solar energy. Nobody is saying . . .

... that the plant thought about it. I agree that we do live according to the principle that one hand washes the other. But still, much behavior can't be explained by that principle. Think of soldiers who throw themselves on an exploding grenade so that their comrades will survive. There are accounts of incidents like that from plenty of wars.

Yes, yes.

Things like that happen. And you can't explain them by your reciprocal altruism, can you?

Well, I think you have to call it a mistake. If you look at the recruiting propaganda, say, in the First World War, on both sides you would see huge social pressure. In Britain during the First World War there was a time when young men who weren't in uniform were handed a white feather, which meant "coward." There's this immense social pressure, and "Your country needs you." Anybody who doesn't join up is a coward, a traitor. Girls won't go out with him. And so the rational strategy in war would be not to go, to be a pacifist. But then other considerations, which may have their own Darwinian advantage, like wanting to be popular, wanting to be especially popular with women, wanting not to be looked down upon, not to be seen as a coward, not to be despised, wanting to conform with social norms — these are all pressures that have their impact for Darwinian reasons.

In Germany at that time soldiers were thrilled to go to war. I don't know whether that was the case in Britain, too.

It most certainly was. In the early days of the war it was exactly as you say. The poet Rupert Brooke died before he had time to realize how awful war was. And so we have some poems by Brooke such as "Now, God be thanked Who has matched us with His hour."

Exactly. Those young men were clearly prepared to sacrifice themselves. I doubt that this can be explained solely by reputation concerns.

I think that's right. And I do agree we have a lot of explaining to do with respect to human super-niceness. But it's not that much more

mysterious than explaining other very peculiarly human characteristics like music and philosophy and mathematics. I don't think
there's any easily understood way they benefit survival.

> I think we've learned a lot about the nature of altruism during these
> last thirty years. And I find it encouraging that we are beginning to
> see in what wonderful ways evolution could use selfish-gene impulses
> that make us care for others.

Yes. I think we have to do our Darwinism in a very sophisticated
way. Here's a very crude and simple example I'm fond of: People
often wonder why insects fly towards light. One hypothesis is that
in nature there is never a light source that's close. Rather, light
always comes from a celestial object, like the sun or the moon.
Because the insect can rely on the assumption that sources of light
are in optical infinity, it uses the light as a compass. It maintains a
fixed angle relative to the rays of light, which in the case of the sun
means it just carries on in a straight line, because the light rays are
parallel. But in the case of a candle, the rays are not parallel and it
spirals in. The point of this parable is that it's the wrong question
to ask what is the survival value of self-burning behavior in moths.
It's not self-burning behavior; it's a mistake, caused by the fact that
moths are now living in an artificial environment where there are
man-made lights.

> Would you explain what you call anti-Darwinian dumbness, or super-
> niceness, in a similar way?

It's easy to understand in Darwinian terms why people are nice to
their kin. But natural selection never favors the cognitive awareness of what you are doing. It doesn't favor animals that know
who their kin are in a cognitive way; it favors rules of thumb. So
in the case of a bird feeding its babies in a nest, the rule of thumb
is if you see a red squawking thing—the red gape of a squawking
mouth—you get food into it. The nervous system obeys that
rule, and in nature that always means: feed your young. But not
quite always: Cuckoos are a simple demonstration that mistakes
happen. [The cuckoo is known to lay her eggs in other species'
nests, letting other birds raise her young.] The mistake seems

painfully obvious—I mean, the nestling cuckoo could be ten times as big as the foster-parent sparrow that's feeding it. But mistakes happen because rules of thumb are not cognizant or wise; they are just rules of thumb. And the rule of thumb still goes on, it still works, it still happens, even though we as biologists can see that this is a cuckoo. And so, for humans, the rule of thumb is: Be nice to kin. At a time when our ancestors lived in small villages, fellow clan members would have been kin, so the rule of thumb to be nice to everybody you meet would mean that, incidentally, you'd happen to embrace kin and usually only kin. Moreover, everyone you meet, you are going to meet again and again and again throughout your life, and therefore there was a good chance that they would repay a favor. But nowadays we don't live in small villages, we live in big cities. So when we meet somebody at random, the rule of thumb that says be nice to everybody you meet has now become a mistake from a Darwinian point of view. But it still works.

Who says that early humans lived only in the circle of their kin? In tribal societies that have survived into our time, anthropologists have found larger group structures. But hunter-gatherers stand up for one another nevertheless. Analogous findings have recently been made even with chimpanzees: The apes in the wild raise orphans that are not related to them. Interestingly, they do so particularly when prowling leopards threaten their community. You can't explain this by selection for kin, can you? It seems instead a kind of mutualism. Each individual has an interest in the community not becoming too weak. Otherwise, each member's own life would ultimately be in danger, too.

Yes—that's also respectable Darwinian theory. And if the group is fairly small and has some permanency to it—so that individuals can expect to see the same individuals again and again—then fortunately kinship and mutualism work in the same direction.

I think that Darwinism is perfectly apt to explain niceness. But I wonder if, in order to do so, our understanding of Darwinism must not so narrowly focus on genes. Shouldn't we take the effects of the group and of the environment much more into account?

I agree, provided that you regard the group as part of the environment in which the selection of genes goes on. So those genes survive best which exploit the fact that they are living in a group.

The American anthropologist Sarah Hrdy has proposed a theory that shares some of your ideas [see my conversation with Sarah Hrdy, page 217]. She points out that because of the high investment required to raise a human child, two parents on their own in nature could never procure enough food to meet the needs of their offspring for a secure start in life. Plus, our youth lasts very long, because the human brain learns an extraordinary amount and so matures extremely slowly, so they need to care for their children for ten if not fifteen years. Therefore, humans could only successfully reproduce and raise children if a community supported them, so only the nice ones—the people whom the community would support—would have a chance. As a result, Hrdy argues, humans had to become the nicest of all apes before they could also evolve into the smartest.

I find this really convincing.

But if Hrdy is right, our tendency toward cooperation is far more than a mere by-product or a "mistake." Nature has virtually built us to share with others. In *The Selfish Gene* you famously wrote, "Let us try to teach generosity and altruism, because we are born selfish." Do you still think so?

That's one of the sentences I would change. It goes against what I was saying, that it's genes that are selfish, not individuals.

And perhaps *The Selfish Gene* is a somewhat misleading title, because it does not stress the importance of the environment.

Yes, I think the title *The Selfish Gene* may be unfortunate. Maybe the book should have been called *The Altruistic Individual*.

In fact, most humans seem to be programmed to feel pleasure when they help or share, as recent neurological studies have shown. When the subjects voluntarily give something away to others, the same brain circuits are activated that elevate our mood when we eat chocolate or have sex.

What about phenomena like Wikipedia and open-source software?

Studies have shown that these kinds of collaborative systems are driven by a sensation of pleasurable sharing rather than concerns about reputation.

I would never have believed that Wikipedia would work. I'm dumbfounded that it works, but apparently it does—people give a lot of time and go to a lot of trouble to make it work. I would actually like to know more about it. It's clearly not for money. Are you sure it's not for reputation? I'm inclined to think you're right.

Yes, I am. When asked about their motivations, most Wikipedians said that they contribute either for the pleasure of writing or for the joy of sharing their knowledge with others. And such answers strongly correlate with the amount of time people actually invest in the project.

I can remember I was once addicted to computer programming, and I would really, really want other people to use what I did, so I would always write a program in such a way that others could use it. I would make use of friendly pull-down menus and things like that, which was not necessary to the particular purpose that I had in mind, and I was really sad if other people didn't adopt my programs.

How should we encourage people's pleasure in sharing?

First, we should take courage from the fact that despite what we might think about Darwinian heritage, at least some people are super-nice. That should drive us, because we have a goal that seems to be attainable. And how to do it educationally? Tell people, tell children, to think it out for themselves. What kind of a society would you like to live in? One in which people help each other, go to each other's aid when they're in distress? Or would you rather live in a society where everybody's out for themselves—a dog-eat-dog society? Encourage the sense of empathy. Ask the sort of question like, "How would you feel if it were you?" I think children are quite likely to get that. They do have a heightened sense of fairness and unfairness. One of the favorite words any child will use is "Unfair!" when they feel that somebody else is getting a better deal than they are. We have to make allowance for that. They even say that when it's inanimate nature that's being unfair. "Unfair! It rained on my birthday, but it didn't rain on her birthday!"

> Do you think we should make allowance for our sense of fairness even when, as you point out, it's so often irrational? Indeed, we do not even mind hardships as long as everybody else is struggling with them as well.

And we don't like feeling that somebody else is getting away with something. In a country like Britain, where people pretty much will pay their taxes, I don't mind paying taxes. But if I lived in a country where just about everybody gets away with not paying their taxes, I would then feel very resentful of paying my taxes. And so, I think it's difficult to foster super-niceness. You can only foster a kind of slightly limited grudging niceness, because most people would be happy to be nice as long as they feel that not too many other people are exploiting it.

> This is why the belief in the goodwill of fellow human beings matters if you want to spread fairness and altruism. Swiss scientists tested whether their students were willing at the time of their enrollment to make an anonymous contribution to a fund for underprivileged foreign students. Half of them were informed that most of their peers had made the voluntary payment; the others were told the opposite. The first group displayed a far greater willingness to donate.

Many studies bear this out, and I feel it very strongly. Introspectively, I can empathize with that.

> There might be a deep evolutionary reason for our sense of fairness. In nature, only relative—not absolute—advantages matter. As soon as I am better off than others in my environment, I will be successful.

Yes. There's a rather cynical aphorism by Gore Vidal: "It is not enough to succeed. Others must fail."

Would you concede that religions have historical merits in spreading altruism? Even if a secular moral philosophy might well meet our needs today, I would doubt that we would have gotten so far without religion.

You're possibly right. What impresses me is the way moral standards change over history, and become, in my view, better very fast. I mean, now we have smart missiles and smart bombs that are, with enormous expense, designed to home in on military targets, and avoid civilian targets. Contrast that with the carpet bombing of Dresden, of Coventry, of Hiroshima. So that's a mere seventy years or so. And back then the same anti-Semitic sentiment that Hitler implemented in political terms, you could hear in any cocktail party conversation in England.

Maybe people are beginning to realize that we are interdependent to an unprecedented degree. We'll have no choice but to learn to cooperate and share across borders. In a rapidly shrinking world, it will be more and more risky to follow only the principle of self-interest. Increasingly, looking out for the welfare of others might be a recipe for success—and a necessity of life.

I agree.

In the Hall of Illusions

Neuroscientist

V. S. Ramachandran

on consciousness

◆

SOMETIMES A SINGLE REMARK CAN HAVE unintended ramifications for years to come. So it was for me with something Vilayanur Ramachandran said in the course of our conversation. At one point in the occasionally meandering exchange, he suddenly came to the subject of happiness. And, as if he couldn't hold back his thoughts for another moment, he asserted, "Everyone talks about it, but no one knows what it is." That observation took root in me—and ultimately inspired my book *The Science of Happiness*.

My conversation with Ramachandran took place in the spring of 2000, long before the others in this book. We spoke in his office at the University of California, San Diego, surrounded by brain models, bones, antique and modern telescopes, and statues of Hindu deities—for Ramachandran is active not only as a neuroscientist, but also, in his spare time, as a collector of sculptures from his native India, an astronomer, and a paleontologist. A dinosaur discovered in the Gobi Desert was even named after him.

Born in 1951 in the South Indian state of Tamil Nadu, Ramachandran initially studied medicine, intending to become a medical doctor. He later earned a PhD in experimental psychology and neuroscience at Cambridge; it was in that subject that he ultimately became a professor at the University of California, San Diego. Ramachandran has made a name for himself with his varied interests, his bold theories of human consciousness, and his unconventional methods. While other neuroscientists spend millions on their experiments and perform expensive computed tomography scans on dozens of test subjects, he uses quite simple materials. Sometimes all he needs is a mirror, a wooden box, or a cotton swab in order to achieve spectacular results.

....

◆

Professor Ramachandran, do you have something against technology?

I have nothing at all against fancy equipment. We need it and use it at times. But personally, I do research because I find it fun. And high-tech science seems less gratifying to me. The greater the distance between the raw data and the conclusion drawn from an experiment, the more boring it is. You know, I have a short attention span and can never stick with one thing for long. That's why I prefer simple experiments: They go faster. Luckily, I studied medicine in India. There you had to fall back on your intuition and very simple tests in order to make a diagnosis. And if that didn't work, we just had to come up with something.

How do your colleagues feel about your speedy methods?

The appearance that everything is so effortless makes a few of them do a double take. But most people find our work intriguing and—often enough—important.

Still, it's not totally uncontroversial. Some of your colleagues accuse you of occasionally letting your imagination run away with you—even

though *Newsweek* once named you one of the hundred most prominent people to watch in the new century.

That's only a problem if you only engage in speculation. If a scientist has "paid his dues" and made solid contributions in the past, he or she is allowed the luxury of some speculative adventures—for example, the whole thing with the God module. In reality, some journalists went overboard at the time. They wrote that we had discovered God in the human brain. But we never made that claim. We merely pointed out the possibility that humans might be programmed to believe in the supernatural. Of course, that doesn't disprove the existence of God. Anyway, at that time a man had appeared in my lab with a huge bejeweled cross and told me about his conversations with God. He said he had understood the true meaning of the cosmos.

A madman?

No, not in the least. Based on previous work by Norm Geschwind, we conjectured that a region in this man's temporal lobes, a cerebral center behind the ears, was much more active than normal. That's often true of epileptics. His religious experiences might have had to do with this heightened activity. At first we thought that everything he sees and hears might put him in a state of such extreme excitement that he believes he is having an encounter with the beyond. But that couldn't be the case. We hooked him up to a lie detector, along with a few other patients with overactive temporal lobes. Then we said words to them, like "house," "love," "murder," "God." We also showed them pictures. The lie detector measures unconscious emotional arousal. Typically, sex and violence stimulate the most intense responses. But these patients had almost no reaction whatsoever to those words and pictures. However, they were all the more susceptible to religious prompts.

Those effects have actually been known for a long time. People who suffer from overactive temporal lobes—for example, during an epileptic seizure—report mystical experiences. Some neurologists have even attempted to explain famous experiences of religious revelation, such as those of Saint Paul and Saint Teresa, as cases of temporal lobe epilepsy.

That's right. But with the lie detector we were able to show that that brain region apparently responds particularly strongly to religious ideas. And with a new technology, transcranial magnetic stimulation, the temporal lobes of healthy people can now be stimulated too—afterward, they too report encounters with God. From this we concluded that evolution might have equipped the human brain with special circuits for spiritual experiences. That would explain why all peoples have a religion. I should point out that these are highly speculative ideas—and I'm not saying that as a legal disclaimer clause. Any professional scientist has to make it clear in his writings when he is sure of his results and ideas and when he's skating on thin ice. Except on rare occasions, I always try to make it clear—but the press often leaves out the qualifying remarks.

Have you yourself had anything like religious experiences?

When I listen to Indian classical music or look at the moons of Jupiter through my telescope, I sometimes feel at one with the cosmos. The separation of my observing self from what I'm observing seems to dissolve. I don't know whether that's the same sort of experience as those reported by mystics and saints. It definitely seems to me the closest I've come to a mystical experience.

How did the Church react to your theories?

It remained surprisingly calm. Reporters even asked the Bishop of Oxford about the God module. He said he wouldn't be surprised if God had equipped us with something of the sort. And the Brahmans in my country saw it much the same way.

But it's far from clear whether there's something like modules in the brain. According to that theory, the brain, like a computer, consists of various circuits responsible for various tasks. But the circuits between the neurons are very densely interconnected, which makes me skeptical about whether each circuit can be viewed in isolation.

I'm fascinated by the idea that there might be such modules. Of course we won't find them for everything our mind does. But for certain functions, specialized circuits might well have evolved. A good example is our sense of humor. People burst out laughing when specific brain regions near their frontal lobe are stimulated. Now it remains to be explained why a function like laughter developed in the first place. I think it has a lot to do with relief. We find something funny when it doesn't turn out quite as bad as we thought. That's how slapstick works: A man slips on a banana peel, falls in the road—you're worried. The man stands up, wipes the peel off his face—it's all right, you're glad. Maybe laughter originated among our ancestors as an all-clear signal. Do you know, by the way, how the existence of the humor module can be proven?

No idea.

Very simple: You measure the brain activity of an Englishman and subtract from it that of a German. There you have it.

Because you suspect a genetic difference in German brains? Or is there some other reason?

I think it's probably due to culture. . . .

Or maybe it's because Germans don't find your jokes as funny?

That's possible. But I gave a talk in Oxford and had everyone rolling in the aisles. Afterward, a man approached me. With a German accent he first praised my lecture and then informed me that Germans are not really so humorless after all—which just went to show, of course, how right I was.

Anyway—you've made the bold remark that, thanks to neurology, humanity will finally be able to solve some of the ancient problems of philosophy. Were you being serious about that?

That's true to some extent. For more than two thousand years, philosophers have been asking the same questions: What is the self? What happens after death? What is art? They don't get anywhere because they just ruminate. Experiments are what advance our knowledge—because as soon as we can grasp what is happening in the brain during perception, thinking, and feeling, we will also reach a deeper understanding of consciousness.

Many people doubt that human beings are even capable of understanding our own brain. In the head of each and every one of us there are more connections between neurons than there are stars in the entire Milky Way—no other object in nature is anywhere near as complicated. What makes you so optimistic?

There's a lot we might never figure out. But we've made enormous progress. Take vision, for example. A few decades ago, how the brain processes color images was still a complete mystery. Today, color vision is 70, maybe 80 percent understood. Or take the connection between brain and body, which we're investigating in my lab. In the next few decades, we will enter a golden age of mind-body medicine. This will be directly based on neurology and psychophysics.

What can we expect?

We've encountered women with false pregnancy, for example, a rare disease. Women with an extreme desire to have a baby feel nauseous, and their belly and breasts swell so much that they themselves believe they are pregnant. Ultimately they have contractions. But there's no baby on the way—it's all an illusion. Sometimes these symptoms even afflict men whose wives are pregnant. A hormone—prolactin—may turn out to be responsible for this phantom pregnancy in both sexes. The release of prolactin can apparently be caused solely by the powerful fantasy of having a baby.

Remarkable—only it's hard for me to see how the understanding of such an exotic ailment can advance medicine as a whole.

False pregnancies are only a particularly extreme example of the effects of imagination on the body. In a similar vein, it has been claimed that the power of suggestion can sometimes cause warts to vanish. And scientists will study what happens in the brain during hypnosis and yoga. Western medicine has neglected such phenomena for too long.

You became well known when you amputated phantom limbs with mirrors. How does one amputate a body part that exists only in the patient's imagination?

After a real amputation, the brain stops receiving signals from the lost limb. Noticing only that something is wrong, the brain reacts with sensations of pain, from which patients often suffer for years. We sometimes rid them of the pain by having them place their limbs into a wooden box with a mirror. If the left arm is amputated, the mirror is positioned so that it shows the intact right arm on the left side of the body. That makes the brain believe that the amputated body part is there again and everything is all right, eliminating the trigger for the pain. Amazingly, this illusion is enough to diminish the pain in some patients. Our mirror feedback procedure has also been used with even more remarkable success for other chronic pain conditions, such as complex regional pain syndrome, in which a trivial injury leads to severe incurable pain.

There is a sense in which the pain too was an illusion—the arm that seemed to hurt hadn't existed for a long time.

Exactly. We've dispelled one illusion with another.

Sometimes illusions can be quite pleasurable: You've described the case of a man who claimed to have orgasms in his amputated foot when he had sexual intercourse. It felt to him like an enlarged penis.

Patients don't make things like that up. The explanation for that experience is to be found in the organization of the cortex: An area is assigned to each body part, and on that map the genitals are next to the feet. After the amputation, stimuli were able to jump from one area to the other.

That proximity would explain why people play footsy. And why there are foot fetishists in droves but barely any arm or nose fetishists.

Of course. And after a genital amputation, men can even experience phantom erections. Yet we still haven't found any patient whose phantom penis was stimulated by rubbing his foot—odd, right?

But how much do discoveries like that tell us about the possibilities of an average brain? In India there's a yoga-like tradition that attempts to bring together sexual and spiritual life through practical exercises:

Tantra. Maybe Tantric exercises use connections in the healthy brain similar to the ones that produce strange illusions for your patients.

We don't know what happens in the brain during practices like those taught in yoga and Tantra. I definitely don't believe it's all hokum. The way we experience our bodies is constantly changing. And we can also play with this body perception. Most people can easily have the experience of merging with a table.

Can you demonstrate that?

Place your left hand under the table. I'll now stroke your left hand with my one hand and the table with the other, making both movements in exactly the same rhythm. Look only at the table, not under it. After a while, you should feel my strokes on the surface of the table.

I don't know. It feels weird. But the sensation isn't coming from the table or from my left hand, but from somewhere in between.

It takes time. The sensation is due to the fact that I'm touching your hand and the table in perfect synchrony. This is so improbable that your brain doesn't entertain this possibility. Instead it perceives the table as part of your body, even though you of course know how absurd that is. The experiment shows how arbitrary and malleable our image of our own body is. Our body, as we perceive it, is a phantom.

The objection could be raised that your experiment is highly artificial.

But it's not. Experiences like that are quite commonplace. Certain centers in the parietal lobes of the cerebrum are constantly determining what we experience as part of ourselves. By way of these mechanisms, lovers feel as if they are fusing together. And parents often say that they live through their children and experience their children's pain as their own. This is far more than a metaphor. And here in California you might even think that the drivers become one with their cars.

Anyway—two and a half thousand years ago, India's sacred Vedic scriptures already warned against being too invested in the idea of the body as the seat of the soul. After all, we don't identify that strongly with our shadow either.

In everyday life the images people have of their bodies often seem really absurd. Anorexic women seriously believe they're too fat.

Among certain stroke patients, those effects are even more extreme. We had a woman who was paralyzed on one side but knew nothing about it due to her brain damage—even though her mind was completely lucid. I asked her to touch my nose with her left hand. She couldn't do it because her arm was paralyzed. "Mrs. D., are you touching my nose?" I asked her. "Of course," she replied indignantly. "I see my fingers less than two inches from your face." She really saw the fingers: Her brain had fabricated this perception.

Why?

The two halves of the brain have different tasks: The left half is the storyteller. Among other things, it is constantly making up theories about the world. That's useful because we often don't have enough information to make decisions. So the left hemisphere of the brain simply fills in the rest, constructing a story that seems coherent. The right half, however, checks these ideas against reality. In the case of Mrs. D., some of those brain regions on the right side were damaged. That's why she couldn't tell that there's a difference between what she intends to do and what she does. In her world, she realizes her wishes.

But couldn't she know that her fabrications are logically impossible?

When in doubt, logic is of little value; other authorities in the brain are much stronger. Just look at some scientist colleagues of mine: brilliant biochemists who believe in spiritual healers. And 90 percent of the American population considers its intelligence above average. How is that possible? Due to flattery from the left half of the brain. Studies show that people are chronically overoptimistic. Slightly depressive people are often more realistic—they pay a high price for that.

The question is whether different functions of the brain are prone to illusion to different degrees. At least with our elementary sense perception of the external world—such as vision—we might stay closer to reality than in our thoughts and feelings.

That strikes me as unlikely. For example, we've examined patients who constantly hallucinate comic book scenes. There's interference with the transmission of signals from their eyes to the visual centers of the brain. That causes parts of their visual field to remain empty. And the brain immediately fills this vacuum with its imaginary perceptions.

Amazing. But why don't all people who have gone blind have that experience? And why comic books?

We'd like to know that too. Probably hallucinations of that sort are far more common than previously thought; people don't like to talk about them because they're afraid others will think they're insane. The brain apparently tends to draw material from childhood impressions—hence comic books. It also might have to do with the fact that comics are simpler than images from real life. They don't move, and they have no depth. What we observe in these cases are of course extremes. But the same mechanisms are at work in all of us. We just don't notice it, because our brain seems to function so smoothly. But in reality we're hallucinating all the time. What do you see? [Makes a cross with his index fingers]

Two fingers crossed.

Exactly. Yet your retina received only the image of a finger and two finger halves behind it. Your brain supplied the notion that it's two whole fingers. We know much less than we think. More than 90 percent of what we believe we're perceiving we're only assuming. These assumptions are part of our worldview, which our brains construct and palm off on us as reality. And these illusions are useful. If our distant ancestors had glimpsed only yellow and black stripes in the bushes, they might not have had much chance to figure out what they were. They did well to see a tiger and run.

How do you investigate the way the brain produces an image like that?

We look at the patients. If we know what parts of the brain are defective and what these people are suffering from, we can infer from that how the brain works—as when a mechanic checks an unfamiliar engine. That's exactly why I find the idea of modules in the brain so attractive. In recent years we've learned that the various parts of the brain don't function like an army, in which all the soldiers march in lockstep in the same direction. Instead, a few dozen circuits are active in the brain, working together to a varying degree. We call these modules. What some of those modules do we can sense; the modules for semantic aspects of language, for example, are connected to consciousness. But most of the modules perform their tasks without our noticing. We call them the zombies.

Zombies?

Francis Crick, a codiscoverer of the structure of DNA who later devoted himself to the mysteries of our consciousness, coined this term for brain modules. It's really quite fitting: Such a circuit is in control when you reach for an object, grasp it, and aim it somewhere. You can post a letter through a slot without any involvement at all from consciousness.

People usually know when they're sending a letter.

But they can slide the letter into a slot. They can perform a simple action like that completely unconsciously. The idea that we do things as a unified individual is helpful for us as we go about our daily lives. But not much in the brain is consistent with it. As far as we can tell, there is no single place in our head that is firmly connected with our being a unified person.

A self.

If we experience something like that, it's probably an illusion. Even our image of our own body is highly unstable. And the vast majority of our mental processes are handled by the zombies, completely unconsciously. What drives us is not a self—but a hodgepodge of processes inside the skull.

The first Western thinker who expressed that notion was the Enlightenment philosopher David Hume. But his theory that the self was only an illusion never gained wide acceptance—for the idea of a self is still quite practical. Life is simpler if I assume that my hand, my wife, my children, and my bicycle belong to me.

You said it: Without the idea of a self, we would be paralyzed pretty quickly. Still, the self is imaginary—a fleeting, often tragic construct of the brain. That's just what the sages claimed in India thousands of years ago.

According to them, human beings suffer from the false belief that they possess an individual soul. This illusion is known as maya. The path to enlightenment consists in seeing through the facade of the self and recognizing your oneness with the cosmos.

That's right. For a long time, I regarded all those traditions as nonsense. I come from a Western-oriented family; my father worked for the UN. And Western medical training certainly teaches you not to give too much credence to mystical ideas. Only after fifteen years of neurological research have I become convinced that in a certain sense the doctrine of maya might have a lot going for it.

Why did you become a neuroscientist?

I already began doing experiments as an adolescent. At one point I wrote an essay on how vision with both eyes works; an uncle of mine sent it to the famous London magazine *Nature*, and they actually published it. That gave me a huge boost of self-confidence, of course. Later I was vacillating between embryology and neurology. The way a fertilized egg turns into a person is at least as interesting as consciousness. But I found neuroscience even more exciting because it deals with your own perception, your own mind. You can look inside yourself.

Does introspection advance your research?

Sometimes it does—but sometimes all these questions about the self only confuse me.

The fact that Eastern philosophies came to conclusions similar to those of neuroscience more than two thousand years ago is astonishing. I wonder why ancient Indian thinkers in particular were so successful in anticipating certain modern insights—even though that tradition wasn't based on experimental science.

In our culture people always had a penchant for speculation. That's why we already had a symbol for zero and place-value notation when people in the Western world were still trying to calculate with those ridiculous Roman letters. But of course, speculations on the nature of consciousness don't yet amount to a scientific theory. The Vedic scriptures also contained reflections on the smallest building blocks of matter before the Greek Democritus philosophized about atoms. Certainly we should pay tribute to such pioneers. But you can't say that our Indian scholars—or Democritus, for that matter—discovered atoms. Their ideas were still too vague. The same goes for consciousness. Indian philosophy primarily stresses one aspect—how deceptive our perceptions are. But that's far from everything that remains to be found out about brain and mind.

Hindus and Buddhists also believe in reincarnation. As a scientist, you could hardly agree with that.

Are you sure? I don't expect to come back as a dog or pig—that's nonsense. But things look different when, instead of understanding consciousness as bound to particular manifestations of matter, you think of it as information—which it is. Then you're reborn every few years. The atoms in my brain are constantly being replaced, but I'm still here. In that sense, I've been reincarnated roughly fifty times, and I'm doing splendidly.

That ends with your death.

At that point, the information from my brain is passed on to my children and other people. Indian philosophy offers a metaphor for that: There's a single divine light that shines through each of us, but individuals are only the windows through which it shines. When someone dies, his window is closed. But the light keeps shining through all the other windows.

But everything that constitutes you as a person is gone at that point.

True. But don't forget: Only the smallest amount of the content of our consciousness is private. More than 98 percent of what we think, feel, and experience we've absorbed from our culture. I know what a table is, what naked men look like, who Einstein was. You know all that too. Only 2 percent, say, of my consciousness relates to my life story, my daughter, my son. And that 2 percent is all that passes away. The rest lives on.

Many of your colleagues seem to have come around recently to the belief that Eastern thinking can enrich neuroscience. The Dalai Lama, whose Tibetan Buddhism is itself based on Indian philosophy, has given talks at the annual meeting of the American Society for Neuroscience—thirty thousand scientists from all over the world attend this conference. Do you think that his ideas contribute to scientific progress?

Yoga, Tantra, and meditation will provide certain insights. But everyday life will benefit from those practices far more than science will. Western societies are overwhelmed by anxiety. That has to do with the fact that people here aren't accustomed to confronting the fundamental questions of their existence: What constitutes a fulfilling life, or, what does "happiness" mean? Everyone talks about it, but no one knows what it is. In the Eastern tradition, however, it's a completely normal part of life to ask questions like that. We have to cope with the cacophony and dissonance of contemporary society.

What is happiness for you?

Passion and devotion. Unfortunately anyone who pursues something too passionately arouses suspicion. To be cool is the ideal. At most there's a passion for mediocrity. Children who deviate from it are ostracized in school. And why is Tom Hanks considered such a great actor? Because he's an average guy. Take a dozen men, put them in a blender, and out comes Tom Hanks. Back when I lived in India, we found people like that deadly boring. The people we admired were eccentrics: those who engage in idiosyncratic pursuits, just because it fascinates them. Those people have a key to happiness: In their devotion they forget their small selves and re-alize that they're part of the great drama of life. By the way, we never would have left the caves if our ancestors had been so devoid of all passion and curiosity.

> That lack of passion might change. After all, the way people expe-rience the world today won't necessarily last forever. It's possible that our descendants in the distant future will attain higher states of consciousness.

Yes, that's quite possible. The complexity of the brain is immense, and despite all our progress, we've understood only the tiniest portion of it. It's therefore possible that the brain can transcend its own limits and expand consciousness in an as-yet unsuspected direction. Maybe this will happen through the natural course of evolution, or maybe through meditation, drugs, and enlighten-ment—or through the installation of electrodes in the brain. My colleagues often ask me, "Rama, isn't all this highly implausible? Why speculate on it?" I reply, "But you don't know that." If they had encountered the first specimens of our prehistoric ancestors *Homo habilis* in the African savanna, they never would have guessed that this species would develop into a creature that writes sym-phonies, understands the cosmos, and attempts to fathom its own consciousness.

The Recalcitrant Zebra

Physiologist and geographer

Jared Diamond

on chance and necessity in history

◆

JARED DIAMOND LIVES IN A RIVER VALLEY above Los Angeles; old poplars, oaks, and camellias surround his house. The scientist has just returned from his morning bird watching and invites me into a large reception room. An open grand piano evokes the atmosphere of an elegant villa, but the masks, statues, and a wooden fish brought back from the South Seas are more reminiscent of the home of a tropic explorer of a bygone era. And as soon as Diamond sits down on a bench covered with a woven carpet and begins to speak, you feel as if you've been transported into the jungle of Papua New Guinea.

No fewer than twenty-four expeditions have taken my host there. From his experiences on his journeys emerged the international bestsellers *Guns, Germs, and Steel* and *Collapse,* in which Diamond attempts to explain the major trends of world history—unusual enough for a scientist who for the longest stretch of his career had a reputation as the leading expert on gall bladders. Diamond, born in 1937 in Boston, studied medicine at Harvard University and Cambridge and in 1966 began a professorship in the Department of Physiology at UCLA. At the same time, however, he was investigating the evolution of birds; later, he branched out even further to study the development of human civilizations. Since 2002 he has been professor of geography at UCLA. A few days after our conversation he will again set off for several weeks of field research on the island of Borneo.

◆

Professor Diamond, over the years you've practiced more than half a dozen occupations: physiologist, ornithologist, geographer, conservationist, anthropologist, evolutionary biologist, paleontologist, historian. You've even written a book entitled *Why Is Sex Fun?* What are you actually?

In our childhood we're all interested in a lot of things. In my case, I had already begun watching birds as a child. But I also loved history as well as Latin, Greek, and modern languages. And then those of us who go on to higher education, we're told we have to specialize. And we put aside all these interests. When I finally got my PhD, it gradually dawned on me that I was expected to devote the rest of my life exclusively to the gall bladder. The idea of screening out everything but gall bladders struck me as a horrible prospect.

And how did you escape that fate?

I traveled with a friend to the Amazon. In the jungle there wasn't much to do. So we watched birds—and saw so much that we wrote two papers right after our return to the United States.

What do you find so fascinating about birds?

At the age of seven, watching the sparrows outside my parents' bedroom window, I literally fell in love at first sight. Birds can fly from one place to another. . . .

The age-old longing to be able to fly.

Yes. I also have quick eyesight and good pattern recognition, so I was good at identifying birds by sight. And I'm quite musical, so I'm very good at identifying birds by song. In the jungles of New Guinea, you hear a hundred birds for every bird that you see.

And why did you later visit New Guinea, of all places?

For the sake of adventure. At first, the birds were more an excuse for our expeditions. I entered a world in which everything was new and wonderful.

What did you find the strangest?

What is the strangest? It's a different world. It's just like saying, "In Japan, what is the strangest?" For me, in Japan everything is strange. In New Guinea, immediately after my first arrival, I spent the night in a village. When I woke up, I saw the children outside playing war with bows and arrows. They had their little bows and arrows, and they were shooting not arrows that could damage, but grass spears. But they were playing war, they were shooting arrows at each other and dodging—so that was my first morning.

You were pretty brave. In those days, many of the tribes of New Guinea had not yet had any contact with outsiders, and in the area of your first trip cannibalism had ended only recently.

Yes, in 1959.

Were you ever afraid?

Yeah, I've been afraid. But it helped that in the 1960s I was stupid, unthinking, and reckless. There were lots of dangerous situations. Once we capsized out in the ocean in a canoe with an outboard engine, when the driver went too fast and the boat got swamped.

Fifteen minutes before sunset, we were rescued by a boat that happened to be passing.

Were the people dangerous too?

In a certain sense. The people were dangerous to each other. Ever since I started there in the 1960s, I have only gone into already-contacted areas. I have usually not been at risk from the people themselves, but sometimes I have been.

For example?

A typical mistake would be to go onto someone's land without permission. I knew that you have to ask permission to go onto someone's land, and so I would ask permission of what I thought were the right people. But then another batch of people would turn up, and I'd say, "But I asked permission." Then they'd say, "No, you asked—. This is our land." So that's a mistake. There were also a couple of incidents when I went into areas where there were still dangerous individual people. Once I was walking with a group of New Guineans between two villages, three days apart, three nights apart. We camped out a day and a half from the nearest village, and I was sleeping in my own tent, and the men who were my carriers were sleeping in a tent nearby. At night I heard the sounds of someone walking next to my tent. I thought it was just one of my porters going out to relieve himself—even though he was walking not on the side of my tent next to his tent, which would make sense, but on the far side of my tent, which seems strange. But I didn't think of it, and I dozed off, and then I heard some voices and saw bright lights and the fire flaring up, and I thought, "My porters are just talking at night and they're going to keep me awake." So I shouted out, "Quiet!" And just went to sleep. And only the next morning did I find out that there had been a man who had come into the camp. And only two weeks later did I find out from the nearby missionary that the man was in fact a crazy person who, at a time when this area had been brought under control, did not want to settle down, remained living in the forest, and every now and then killed people. He had recently killed his son, and killed various women in the area. So if he had gotten into my tent or if he had come across me alone, then that would have been dangerous. At the time I didn't realize it.

You were leading a double life in those days: Alongside your adventurous expeditions you pursued a completely ordinary academic career in Los Angeles as a professor of physiology. How did that work?

During the day I would work on gall bladders and in the evening and on weekends I would work on the New Guinea birds. Every summer I spent three months there.

At times, did you find the specialization—having to make a living as a gall bladder expert—painful?

No, because I had made it a condition during my job negotiations that I would maintain a career in New Guinea birds. When I made the demand, the dean replied, "Jared, UCLA Medical School is very interested in the birds of New Guinea." He was a tactful man.

How did your family cope with your expeditions?

As soon as my children were born, my trips to New Guinea were reduced to only five weeks a year. My wife never came with me. She's tolerant.

Did shuttling back and forth between Los Angeles and the jungle ever lead to a feeling of split consciousness?

Not really, because I approached the gall bladder and the birds in the jungle in a very similar way: I gathered huge amounts of data. Here I measured the permeability of membranes to two hundred different chemical compounds; there I investigated the distribution of six hundred bird species. But out of the initial mass of data, patterns gradually emerged. Really, the only difference between the physiological data and the data collected in the jungle is that the gall bladder itself can be experimented on. With birds in the wilderness that's not possible.

There you have to make do with the experiments nature itself has conducted in the course of time.

Exactly. For example, you might wonder why a particular species of bird can live in the area you're investigating but another can't—and draw your conclusions from such comparisons. I've followed the same approach in my studies on the basic patterns of human history. You can't conduct experiments on human societies either, but you can compare the development of different peoples.

Your book *Guns, Germs, and Steel* begins with the sentence: "This book attempts to provide a short history of everybody for the last 13,000 years." Weren't you ever daunted by the enormity of that task?

Not in the least. Of course, I wrote about many subjects in which I initially was not an expert. But I consulted hundreds of scholars, read their papers, and asked them to check my work. If some historians and anthropologists now claim that I don't know what I'm talking about, they're actually saying that the best historians and anthropologists don't know either.

How did your interpretation of history develop? Did you invent the theories yourself, or did you talk it out with other people?

Something of both. A central idea in my book, for example, concerns the expansion of continents: Europe and Asia were historically at an advantage, because their land mass extends along the east-west axis. That allowed methods of agriculture to spread relatively easily within a single climate zone. On the American continent, because of its north-south axis, the obstacles were much greater. The archeologist Richard Yarnell, who specializes in Indians of the southeastern US, pointed that out to me. But eventually it dawned on me that not only America but also Africa and the Indian subcontinent extend from north to south and so have the same disadvantage.

Your work was to put together the experts' puzzle pieces.

Yes, it's primarily a work of synthesis, but there's also new analysis. For example, the chapter on the Anna Karenina principle. . . .

From the first sentence of Leo Tolstoy's novel: "All happy families are alike; each unhappy family is unhappy in its own way." Tolstoy meant that success has many fathers. In a happy family, many factors have to be right: health, sexual attraction of the parents, money, and so on.

If, however, just one of those factors is deficient, the family members will probably be unhappy.

The principle explains why most animals were never domesticated. Only where many favorable influences came together could domestication succeed. I came up with a list of seven things that animals need to be domesticated—if they're lacking any one of these, domestication doesn't succeed.

The question, though, is whether more than ten thousand years later you can identify which among all the possible factors were decisive back then.

That was the problem I grappled with most intensely during the writing of *Guns, Germs, and Steel*. My argument was that the domestication of plants and animals occurred only in certain parts of the world not because of the people but because of the plants and animals themselves—they had characteristics that lent themselves to being domesticated or not being domesticated. It wasn't some difference between Iraqi and German hunter-gatherers that resulted in the Iraqis becoming the first farmers but not the Germans. But how do you really know that the difference was between the plants and animals and not the people? So that was the area where I worked the hardest to come up with test cases and arguments that would show the way. Eventually I was able to identify a whole series of traits that animals themselves have to possess in order to be domesticated. The strongest effort I found of a people trying to domesticate a certain animal is in southern Africa, where European farmers and herders have been working hard to domesticate zebras and giant elands. But none of their experiments have succeeded—it turns out that zebras and giant elands haven't been domesticated because they are difficult to domesticate.

Your version of human history is a reckoning with racism. Did you ever believe that there could be natural differences in the abilities of different peoples?

Of course I did. On some level none of us are completely free of racial prejudice. Not even the most intelligent people. I remember a conversation with a very distinguished Harvard professor who told me after his trip to Australia how primitive the aboriginals there looked—and how primitive they were, in his view. We Americans have a history of this. In 1941 we suffered the attack on Pearl Harbor because no one had thought the Japanese capable of such a brilliant strike on our fleet. This is of practical importance today because, if we Americans think that those people in Iraq and Afghanistan are stupid, we're making a big mistake—racism is a very expensive mistake.

In your view, would it have been foreseeable twelve thousand years ago that Europe would someday spread its culture throughout the whole world?

Actually, it was a triumph of the peoples of Eurasia. True, the Europeans conquered the world. But the inventions that made that possible for them in the first place were not their own. Nothing interesting happened in Europe until the invention of water wheels in the Middle Ages. Everything else—from metallurgy to the alphabet—came from elsewhere. And yes, I believe that the later development was enabled by the east-west axis of the Eurasian continent and its distribution of domesticable plants and animals.

Do you believe that chance plays an important role in history?

Yes and no. Think of the assassination attempt on Hitler on July 20, 1944. If the bomb had been positioned just slightly differently, Hitler would have been dead. The Second World War would have ended earlier; the map of Europe would look different today. But what difference would that make in four hundred years? Not much of one, I think. In any case, I can't think of any contingency that could have resulted in aboriginal Australians conquering the New World. So contingency can play a role for short times or small distances—the longer the time scale, the larger the distance, the less the role of contingency.

For a really long time, European civilization was far from the most highly developed in the world. Into the sixteenth century at least, China was far superior to our continent culturally and technologically.

And it was, unlike Europe, politically unified.

And that led China not only to great successes but also to awful mistakes.

You've written in *Guns, Germs, and Steel* that China weakened itself when it forbade oceangoing shipping. Isn't that a contingency with very long-term consequences? A different emperor would have made a different decision. In any case, it seems that if China had had three hundred more years to develop, things could have turned out differently.

The question of why Europe and not China ended up conquering the world is perhaps the greatest unanswered question of history.

Why do you choose to present your theories in popular books as opposed to specialist publications?

I write things the same way I explain them to myself. When I begin to familiarize myself with a new field, I'm basically a layman myself—like my readers. On top of that, I find my subjects too important to bury them in academic journals. Publishing in an academic journal doesn't do anything to dispel racist ideas.

Historians must not particularly like you, if you flout their conventions.

Every year I get a couple thousand invitations to give talks. But only two have come from history departments: one from the university where my son was studying, the other from my own.

An important difference between science and the humanities is that scholars in the humanities focus more on individual cases, while scientists like you look for general patterns and rules. Do you think our understanding of history would change if historians mastered statistics?

Yes, I do. But the way you put it is going too easy on historians.

What do you fault historians for?

Many are so enamored with their individual case studies that they disregard the larger picture. As a result, history becomes a story that has nothing more to say. Take the American Civil War. Huge tomes are written about individual battles, or even about just the first day of the Battle of Gettysburg. But if you really want to understand that conflict, you have to compare it with other civil wars—with the Finnish Civil War of 1918, with the Irish Civil War, with the wars between Austria and Prussia. Then you can figure out what was different about our civil war and what was typical. But American historians of the American Civil War don't write these books.

> That problem is not unique to history. In all disciplines, academia breeds extreme specialists. Personally, I found that tunnel vision, which an academic career simply requires, horrible.

I agree. It's not that everyone has to study gall bladder physiology and history at the same time. But extreme specialization prevents scholars from developing a perspective. And only that perspective would allow them to come to valid conclusions—instead of just telling stories. Recently, my friend Jim Robinson and his colleagues showed how it can be done differently. There has been a long debate about whether Napoleon's reforms were more to the benefit or to the detriment of Europe. Usually, historians pursue such questions by way of an example, such as the city of Cologne. Robinson, however, compared the development of twenty-nine German cities. In some of them Napoleon had left the rulers intact, while in others he threw the rulers out. And he was actually able to show that the cities in which Napoleon carried out his reforms developed better in the long run.

> Do you believe that the study of history can be of practical value?

We can learn from the past about the bad things we have done in the past and are still doing today. Today we're still deforesting, we're still overfishing—overfishing has big consequences, and it has happened in the past. Those are some of the general things that we can learn. There are also cases where you can learn very specific things from the past. That's why the government of Iceland now has a whole department devoted to land management, studying the rate of sheep stocking and grazing in the Middle Ages, because what you can learn from the number of sheep in the Middle Ages in Iceland is very relevant to the number of sheep today.

But nowadays we're no longer as dependent on agriculture as the past societies you describe. Today overgrazing is the least of Iceland's problems.

Fewer and fewer people are farmers; crops are grown on an ever-smaller proportion of the land. That makes us more vulnerable to ecological adversity than ever before—most of the population is supported by so little farmland. In Germany with its robust soils you might not feel the problem so acutely. But in China the best farmland now yields something like a third less than just a few decades ago, because fertilizers and pesticides have ruined the population of earthworms, which churn and regenerate the soil. It's no accident that Islamist terrorism has flourished in ecologically devastated lands.

But ecological disasters can't explain Osama bin Laden.

The connection is that Osama bin Laden requires a poor, desperate country without hope. Ecological devastation fosters the willingness of the rural population to support or tolerate the terrorists.

Correct me if I'm wrong, but ecological destruction—namely, deforestation—ruined the habitat of Easter Island.

That's right. The society was dependent on wood: for cooking and fuel, for building their huts and canoes, not to mention transporting and erecting their gigantic statues. So six hundred years ago you could have said that the last thing these people are going to do is cut down all their trees, but they did cut down their own trees.

In your book *Collapse* you investigate the fall of civilizations. But can there actually be general explanations for that? The Anna Karenina principle, according to which every unhappy family fails in its own way, seems to forbid it. How likely is it that the pattern of past errors predicts the pattern of present or future errors?

Some things about the past have lessons for the present, but there are also differences. You have to examine these differences—some make our present risk worse, and some make our present risk better. Easter Island was devastated with stone tools; now we have chainsaws. And back then there were twenty thousand people living on the island; today the population of the world is close to seven billion. So a difference between the past and present is that the destruction that happened in the past is happening more rapidly today.

You must have found the conditions on Easter Island depressing.

More tragic than depressing. It's entirely self-contained, so what happened—they did it to themselves.

We imagine that a global ecological disaster will wipe humanity off the face of the earth like the dinosaurs. But in reality it might not even destroy civilization. Only the living conditions would be wretched.

Exactly. Imagine if the whole world were like present-day Somalia or Haiti: Then you have the worst-case scenario. If we continue down the road we're on, using resources the world just doesn't have enough of, we'll be there in thirty years.

In your next book you intend to compare traditional societies with modern ones. What can we learn from people in tribal cultures?

My friends in New Guinea always struck me as relatively self-assured. They're autonomous. They're comfortable doing things by themselves. They're more curious than Americans or Europeans. They're not afraid of being abandoned, being relatively open with their feelings.

Do you mean that in the course of the development of our culture a natural self-confidence got lost?

I trace the differences back to child rearing. That's one area where we could learn from people in tribal societies. Boys and girls in New Guinea grow up with a level of autonomy that can seem shocking to us. For example, I noticed among one tribe that almost all the adults had burn scars. They're from the first years of life. Because parents think that a baby should crawl wherever it wants—even into the fire. When it gets burned, they take it out. Personal freedom simply has its price. But we can also learn a lot from the way tribal societies assess risks and treat their old people.

In what time and place would you want to live if you could have your pick?

I would choose twenty to maybe thirty years ago. I'm not fond of the world of computers and video games. And I loved living in Germany. Of all European countries, Germany is the country I enjoy most. I had a German girlfriend, and if things had worked out differently. . . . But as I've grown up, I've realized that there are things that would bother me, if I were living there, that didn't bother me when living there for a short time. The countries I would consider moving to, realistically, are Britain, Australia, and New Zealand.

And do you have a wish for the future that you're most hopeful will come true?

My deepest concern is that our world's current course is so unsustainable that it has no future. In fifty years, my boys will be the same age I am now, and by that time, I'd like their world to have a far more sustainable economy, which is most important, and I also hope we've achieved rates of consumption that are much more equal around the world. Fifty years from now, I'd like a world for my boys that actually has a future.

Chimps Are Individuals Like Us

Primatologist

Jane Goodall

on our relationship to animals

◆

I HAVE NEVER MET A MORE humble celebrity than Jane Goodall. When we talked in her small hotel room in Munich, the famous animal behavior scientist had just finished an arduous lecture tour through Austria. Next on her agenda was the Swiss premiere of a film about her life. But Goodall, born in 1934, not only spoke with extreme acuity, occasionally displaying a very British self-irony, but also listened with a level of attentiveness I have rarely encountered.

Yet, according to an American study, only one scientist is more widely known than she is—Albert Einstein himself. Goodall would probably object to the mention of his name and hers in the same category. Still, it was a pioneering act comparable with Einstein's conceiving the theory of relativity when she, after completing secretary school, began in 1960 to investigate the lives of chimpanzees in Gombe National Park in East Africa. Never before had anyone spent an extended period of time observing wild apes. And what Goodall has discovered in the more than fifty years since has not only opened up a new understanding of our closest relatives, but has also pointed the way for our relationship with animals in general.

On the second day of our meeting, as we strolled through the nearby park along the Isar River, Goodall suddenly uttered chimp noises so deceptively authentic that passersby turned around, clearly afraid that an escaped animal was roaming the grounds.

◆

Dr. Goodall, is it true that you have difficulties remembering faces?

It took me a long time to realize it. But then Oliver Sacks, the famous neurologist, explained to me that I suffer from a disorder known as prosopagnosia, or face blindness.

If we ran into each other tomorrow on the street, we might not recognize each other.

I think I would recognize you—I don't have it too badly. I have trouble with average faces but I would remember yours. And you would know who I am, wouldn't you?

Maybe not. I actually have the same disorder. The worst experience was picking up my children from day care and recognizing them only by their clothing.

Oliver Sacks still does not recognize his secretary after ten years.

Do you have the same problems with chimpanzee faces?

Yes. I managed by making a conscious effort to memorize the structure of the face, along with things like posture and hair color. Once I know a chimp or a person well, then I know them always.

If one wanted to sum up your life's work in a single sentence, one might say: She gave chimpanzees faces. Previously animals were regarded as interchangeable. You showed that each one has its own personality—like us.

And a mind and emotions.

A long tradition of Western thought sees that quite differently. The French philosopher Descartes, for one, described animals as machines controlled by their instincts. Where did you find the courage as a young woman to disagree?

I was really fortunate in two things. One was my mother's wisdom. She always taught me that if you meet someone who disagrees with you, who thinks differently from you, first thing is to listen to them, to keep your mind open, and if you still think you are as right as can be, then you must stick to your convictions. The other was my teacher. Have you read about him?

Louis Leakey, the famous anthropologist. In 1960 he sent you, accompanied by your mother, to the East African forest to study the chimpanzees. He was hoping to find out about the origins of human beings.

No! My teacher was Rusty, my mutt. He was with me through my whole childhood. He was incredibly intelligent—and really different from all my later dogs. Thanks to Rusty, it never even occurred to me to doubt the minds, personalities, and emotions of animals.

He immunized you against the prevailing views.

At that time, I didn't know that personality, mind, and feelings were supposed to be unique to us. I had done just high school biology, and our teacher really did not know that much about zoology. Then I left school. So why would I possibly know that animals were not supposed to have minds and feelings? All my animals had. If somebody had said to me as a child, "Your dog does not have a personality," I would have just looked at them in amazement and said, "Well, he does."

So ignorance, if it's combined with cleverness, can be a plus at times.

Absolutely.

Then you met Louis Leakey, and he offered you the opportunity to go into the forest to study primate behavior. But he must have known well that you had never seen a university from the inside.

Louis told me afterward—but not at that time—that he wanted someone who had not been to college. He wanted somebody whose mind was unbiased. Had he told me *then* why he chose me, I would immediately have wanted to find out what I was supposed to know. But he did not, you see.

You were part of an experiment, too.

Yes, but I did not know I was. He was very wise.

You were one of three young women Leakey sent to observe apes: He assigned Dian Fossey to the gorillas and Biruté Galdikas to the orangutans. Why only young women?

Because he preferred working with young women. He was not so much at ease with men—maybe it was the competitiveness of another man. He just got on very well with women. And now and then he went too far. It could have been a problem for me, because he was the only person who could make my dream come true, but fortunately we were able to work it out.

So things turned out all right?

Louis had obtained money from a wealthy American for six months of my research, and that was it. I had horrible binoculars. Everything was secondhand. For the first four months my mother was with me—the authorities would not allow a young girl on her own in the forest. We lived in a former army tent. If you wanted to let some air in, you had to tie up the flaps, and then spiders, scorpions, and snakes would creep in. My mother not only put up with all that, but also boosted my morale. In those first days the chimps took one look and ran away—they were frightened by the strange white ape. I could only watch them from afar. My mother kept telling me I was learning more than I realized. But I knew if I did not see something exciting before the money ran out. . . .

You were desperate.

I was. You know, I was living my dream, but all I could think of was that if I didn't see something exciting before the money ran out, that would be the end of the study. And I would have let Louis Leakey down.

What changed the situation?

The encounter with David Greybeard. I had observed this chimp before and could recognize him—I was very lucky that he was so distinctive. He perhaps always had less fear of people, and so he came into my camp several days running and climbed the palm tree there to eat the ripe fruits. And then he found some bananas and took them off the table. I began leaving bananas for him. Soon he let me get close to him and introduced me to his friends in the forest. That's how I came to observe him using tools. We were defined back then as "man the toolmaker." So when I sent Louis Leakey a telegram about it, he famously said, "Now we must redefine 'tool,' redefine 'man,' or accept chimpanzees as humans."

Were you proud of your discovery?

Well, if I had not done it, somebody else would have, I'm sure. I was just lucky to be the first—I mean, wow, what a privilege. I suppose the fact that I stuck to my guns about personality, mind, and emotions probably hastened things.

Soon after that you discovered that chimps make tools by stripping leaves off twigs. And that they hunt and eat smaller monkeys. I learned in your book *In the Shadow of Man* that eventually David Greybeard trusted you enough to touch your hand.

Yes. It was amazing.

Weren't you afraid to touch that huge wild animal?

I didn't see him as a wild animal, but rather as an individual whose trust I had gained.

And then I read somewhere that eventually you wouldn't touch them anymore. Why not?

It suddenly hit me, after I'd brought students, that if we go on touching them, we are changing their behavior. It wouldn't be natural anymore. So we stopped, just like that.

And after doing this work in Gombe National Park, you went to Cambridge?

Yes, I went there to get my PhD, because Louis had said, "Jane, you are going to have to get money for yourself, and you need a degree for that." And it was not really until I got there that I realized that as far as the scientific establishment was concerned, I had done everything wrong. Some of the professors were just horrified. I wasn't even supposed to have given the chimpanzees names. Back then it was standard to number animals in research. And here I was telling them how David Greybeard and his friends played blind man's buff with a blanket they had stolen from us!

Your professors were looking for the typical ape. You, however, were interested in each individual animal—its history, its particularities.

Yes, I think of the different species as collections of individuals. Many of my colleagues had castigated me for telling mere anecdotes. To me, chimpanzees have such different personalities and such different early experiences, and learning plays such an important role in their individual lives, that a collection of anecdotes really gives a good idea as to what they are capable of. I collect anecdotes about rarely observed behavior, and when you have enough you begin to have a much better picture. That's why I love stories.

As a physicist, I'm attracted to the beauty of general laws. But how can you find them if you're not looking for them? The circumstances of each anecdote are different.

Animal behavior research will never be a hard science. Science can be arrogant—it pushes aside a lot of fascinating things because it does not have the tools to study them. In an experiment at Oxford in 2002, two crows were given wire with a hook, to hook up some food. But then one day the hook broke off. To everybody's utter amazement, one of the two, after much frustration, actually bent the end of the wire into a new hook with her beak and foot.

Really?

The skeptical scientists were convinced it was a fluke. But when they were given new straight wires, the female did it again—the male took the reward from her so did not need to make a tool! Since then, there's been a flurry of investigation about bird intelligence. The irony is that parrot owners have always known that their pets could do things like that. But scientists had pooh-poohed the idea of it because the bird brain was not structured like ours.

Aren't you portraying your colleagues as more arrogant than they are? After all, you yourself have influenced a whole generation of animal behavior scientists. And to this day they bring to light amazing things about wild chimps. A recent discovery from the Ivory Coast surprised me. Chimps there adopted orphans, even though they weren't related to them. Even males tenderly cared for the little ones—and chimpanzee fathers normally don't even take care of their own children.

It doesn't surprise me. We observed chimp adoptions too, though mostly among family. And we know from zoos that chimps sometimes rescue each other from drowning—there was one case where a male drowned trying to save an infant. When an infant starts to drown, even a completely unrelated male will jump in. The origin of this selfless behavior may lie in the close, long-term bonds between mothers and their offspring and between brothers and sisters.

I think I learned from your writing that it would take five years for a chimp to grow up.

That's why in the early days we got all the ages wrong. We could not believe that a five-year-old could still be suckling from the mother.

Do you think we can learn anything from chimpanzees?

The way the mothers relate to their children, for example. I noticed that if you have a high-ranking, supportive, affectionate, playful mother, if you're a male you tend to rise fairly high in the hierarchy, and if you're female you tend to be a better mother. And they have so much fun together. They play with their babies, tickle them, laugh with them. And I thought, "I am going to jolly well have fun with my baby, too," and I did.

What did you do with your son while you were observing the animals?

For the times when I couldn't be with him, we built him a big cage—a caged-in veranda, really. He could play in it, and there was always someone with him.

Why did you cage him in?

To protect him from the chimpanzees. We knew that they take human babies.

What do they do with them?

Eat them. Humans eat chimps and chimps eat humans. They're simply primates like us.

In your first book you described chimpanzees as creatures full of caring, motherly love, and intelligence. And then you realized that they can be brutal.

It was a shock. I had thought they were like us, only nicer. We got the first intimation of how brutal they could be when a student witnessed males from one community attacking an individual female from a neighboring social group and leaving her to die of her wounds. Her infant was also killed. Then came the "four-year war," when a group of chimps who, up to that point, had coexisted peacefully, divided, and the smaller group occupied part of the range they once had shared. Then the males of the larger group systematically hunted down and attacked the males and adult females of the smaller group until they were able to reclaim the territory that had been taken over.

When something like that happens among human beings, it's called ethnic cleansing—a horrible phrase.

It was awful to see how similar to us they are. The young males were fascinated by the murders. They wanted to watch the attacks. It was even worse when we saw a female attack another female from her own community, and then steal and eat her infant. She and her daughter between them killed and ate as many as ten infants.

Did those events diminish your sympathy for the chimps?

It horrified me, but, sadly, learning about this brutal behavior made them seem more human than before.

Some of your colleagues have accused you of causing the war yourself by offering bananas too generously, which resulted in fierce competition.

The bananas had nothing to do with it. The group that split off had lost interest in our feeding. Those animals were living on the fruit of the forest again. And there has been similar aggression between communities in almost all places where chimpanzees have been studied.

Even so, did you ever wonder whether you altered the behavior of the animals through your presence?

Well, maybe. But it was so magical to be able to touch them in the early days. It was the payoff for all those months and months when they ran away. It was wonderful. I would not have forgone it. When it became clear that the study could continue into the future, though, we stopped doing that. We didn't want the chimpanzees to get too used to us humans, and we didn't want to infect them with our diseases.

Still, you forged something like friendship with some chimps.

Actually, there's no word for the sort of relationship I had. A dog can be a friend, a chimpanzee can't. They have their friendships with each other, but not with me. It's probably closest to mutual trust, perhaps respect.

It is a sacred law of modern science that researchers are to maintain distance from what they're investigating, for only in that way can they be objective.

That's why I've always felt like an interloper in the scientific world. The theory is that if you have any kind of empathy with animals—in this case, chimpanzees—you cannot be objective. I totally disagree.

But doesn't that make it hard to distinguish observed behavior from your own projections? When people are in love, they see themselves in the eyes of the beloved.

But I was not in love with the chimpanzees. It is a question of self-discipline. You need only to look at my notebooks. Once I observed a chimpanzee mother trying to comfort her baby who had been attacked. The flesh was hanging in shreds from the wound, her arm was broken—and the more the mother cradled the child to soothe it, the louder it screamed in pain. Tears were pouring down my face. But my notes describing the events were absolutely objective. I think empathy and intuition play a really important role in understanding why they're doing what they're doing.

Can we ever truly understand what's going on inside an animal's head? Even people from other cultures, when we don't speak their language, can be tremendously difficult to comprehend. I experienced that in Japan.

And chimps are even more foreign. Yet their genetic structure differs by only just over 1 percent. My professor at Cambridge was fabulous and I'll never forget him. Naive me, not knowing anything about science, I'd written a piece for my thesis that said that Fifi, a chimpanzee, was jealous of her sibling. My doctoral advisor said, "You can't say that because you can't prove it." He suggested that I say, "Fifi behaved in such a way that had she been a human child. . . ." I have used that advice throughout my career, and I teach all my students the same thing. It's very clever, because if you put it that way, nobody can fault you or attack you, and yet you are saying exactly what you mean.

A famous scene in your first book describes chimpanzees going into a sort of ecstasy at the foot of a waterfall deep in the forest.

When chimpanzees hear the roar of the water, sometimes their hair starts rising in excitement. Then they sway rhythmically from one foot to the other, stamping in the water, for up to twenty minutes. Finally they may sit down on a rock and just watch the water.

You interpreted this behavior as reverence for the spectacle of nature and described it as proto-religious. Isn't that a rather bold claim?

Well, I asked myself where those wonderful rhythmic movements came from, and all I can think of is some kind of wonder or awe or primitive feeling—the sort of emotion I feel. And might it be that sort of feeling that could have led to the early animistic religions— you know, this worship of what you didn't understand, the thunder, the lightning, the sun, the water? I think so. The whole time I was in Gombe National Park, I felt this great spiritual power, something that was greater than me—a great mystery.

You felt connected to a higher order.

Yes. But can I know for sure whether the chimps felt the same thing? I can't. Of course I can't.

When we anthropomorphize animals, we run the risk of failing to do them justice.

True. But the matter is more complicated: On the one hand, we perform experiments on animals. We lock chimpanzees in tiny cages in order to better understand our own diseases—including mental disease—because apes are so similar to us. But on the other hand, we're not prepared to admit that there is any emotional or intellectual similarity.

We're protecting ourselves from empathetic feelings—isn't that what we're all doing? Farmers give names to their pets, but they would never give names to the animals they intend to slaughter.

There was a clever study of workers in medical labs. They don't give names to the rats and dogs—the animals they're experimenting on. But some of those people have pets. And when they go home, they declare that their dog is part of the family and understands every word they say! They're smart enough to know that their pets and the laboratory animals can have the same feelings. But the fact that they name the one animal and not the other protects them from that awareness. We make ourselves schizophrenic. It is weird.

In earlier cultures, like ancient Greece, people were apparently troubled by something like guilt about the mistreatment of animals. To assuage this guilt, the slaughter was often accompanied by an offering to the gods.

Yes, and that is typical of us, isn't it? You say grace, you say, "Thank the Lord for what I am about to receive" or something, and you tuck into a chicken, but you don't thank the chicken, whereas the Native Americans do. They thank the animal for its life. They are that much closer to reality. To every life they make a prayer, and they know they have taken a life. We don't.

Do you believe that we're entitled to kill animals?

It depends on what it's for. We should not kill needlessly. If I was starving and my son was starving, and there was nothing to eat but an animal, I am sure I would kill it and eat it.

But we shouldn't make the way we behave toward other creatures dependent on the closeness of our relation to them, should we?

No. It would be at least as hard for me to sacrifice a dog for my son as a chimp. I probably have even more empathy for dogs than for chimps. The key is to stop thinking of animals as machines and accept that they are thinking, feeling individuals. Then we will understand how we demean ourselves when we cause them pain, distress, and suffering.

You've turned your back on research to devote yourself to animal welfare and environmental conservation. Was that hard for you?

Yes. I really, really miss life in the forest, and I really, really miss analyzing the data—and being able to collect it and ask questions and go and look. But I couldn't go back to that because it wouldn't feel right. At a conference honoring the publication of my major scientific work, I had the shocking realization of how great the threats facing chimps are. The forests are disappearing all over Africa, the bushmeat trade is flourishing. . . .

You didn't know that? You had already been living in Africa for twenty-five years at the time.

I was so immersed in my research—not to mention raising a child—that I didn't even notice what was going on around us. And my colleagues weren't talking about it either. I wish more people in my field would grasp that they should be a human being first and a scientist second.

At first you advocated mainly for chimpanzees. Since then, you have taken up the cause of conservation in general and, on top of that, launched initiatives for the rural population in African countries. What gives you the confidence that we can still prevent the destruction of the earth?

Well, I have always loved children. After finding out what is going on in the world I thought, "Gosh, it is getting very, very hard to conserve chimpanzees and forests. What is the point if the next generation is going to go on trashing the planet the way we have?" So it seemed very important to me to work with children. Roots & Shoots began in 1991 in Dar es Salaam, Tanzania, with twelve high school students very concerned about what was going on around them; today we have groups of all ages, kindergarten through university, in more than 130 countries. The members of each group decide for themselves what they want to do for the environment, animals, and their fellow human beings. They can volunteer in a shelter for stray dogs, or clean up a stream, or remove exotic plants from a prairie, or volunteer in a soup kitchen—it's their choice. The point is for them to learn to live in harmony with other people and the natural world. The reason I have the energy to go around the world aged eighty-one is that everywhere there are groups of young people wanting to tell Dr. Jane what they have been doing to make the world a better place.

It's said that you're on the road three hundred days a year to visit your groups and give talks. People worship you. What is life like as an icon?

Well, all I can do is just to think, "Okay, it's happened." I didn't want it to happen. I didn't work for it to happen. I try to think about it as little as possible. Again, going back to my upbringing, in our family there was a lot of teasing, and there was no way you could get too big for your boots because you were brought down and people laughed. You learn to laugh at yourself, and not take yourself too seriously.

The Unity of the World

Physicist

Steven Weinberg

on science and religion

◆

EXCESSIVE RESPECT IS NOT A FEELING that often afflicts me. But after Steven Weinberg agreed to a conversation, I wondered how I was to encounter a man who, perhaps more than any other living physicist, has shaped our conception of the structure and origin of the universe. Born in 1933 in New York City, he taught theoretical physics at Berkeley and MIT, before moving to Harvard in 1973; in 1979 he received the Nobel Prize. And the Nobel laureate excelled not only as a scientist, but also as a natural philosopher and writer. With his 1977 bestseller *The First Three Minutes* about the period of time after the big bang, he inspired a whole generation's enthusiasm for physics—including my own. With his brilliantly written essays on science and religion he causes a stir to this day.

But all my timidity dissipated when I entered the modest room at the University of Texas at Austin; Weinberg has been a member of the physics and astronomy departments there since 1982. He sat in front of a blackboard with mathematical symbols written on it. Without standing up himself, he offered me a seat and immediately began to engage me in a conversation as if we were old friends. During the conversation he repeatedly laughed out loud. All the while his hands played with the golden knob of a walking stick.

◆

Professor Weinberg, is it true that you made the greatest discovery of your life in a red sports car?

Yes, in a Chevrolet Camaro. That was in 1967. I was trying at the time to understand the strong forces that hold the nucleus of the atom together. But I was getting nowhere. My calculations kept pointing to the existence of a particle of zero mass. But that contradicted all experiments. Then it suddenly dawned on me that the massless particle was none other than the photon. . . .

The long-known elementary particle of light.

Exactly. My ideas had been correct; only they applied to a completely different problem than I'd thought. I had been looking for a theory to account for the strong forces in the atomic nucleus and found instead a theory that accounts for both light and so-called weak nuclear forces.

Like a detective following the clues to a crime and in the process solving an entirely different one.

Something like that. And I realized all that while driving through the streets of Boston on my way to work.

Not very safe.

At least I wasn't talking on the phone at the wheel. But that really is a problem: We theoretical physicists are constantly thinking about what we're trying to do—like composers or poets, perhaps. And so I forget where I parked the car or what I actually wanted to buy in the store I just entered.

Your epiphany behind the wheel gave fundamental physics a new direction. It resulted in the so-called standard model, today the widely accepted conception of the structure of matter and the origin of the universe. Did you suspect all this at that moment?

Most of the time you run into a dead end. In that case I sensed that there might be something to my idea. So it was a great pleasure to work out a theory that might be true. But the fact that I was actually right wasn't demonstrated by experiments until six years later. That was the second great pleasure.

In modern physics there are four known fundamental forces that vary in their strength and range: gravity, electromagnetism, strong nuclear force, and weak nuclear force. You and your colleagues figured out how two of those natural forces—the electromagnetic and weak forces, which seemingly had nothing to do with each other—can be traced back to a single fundamental force. And for a fundamental physicist, to unify different phenomena means something like attaining the Holy Grail. Why is that actually?

Because we want to achieve a simpler understanding of nature. And the path to simplicity is unification. Think of Newton, who discovered that the planets follow the same laws as a stone falling to the ground. So there aren't separate natural laws for the heavens and earth, as people had thought up to that point—only gravitation, which applies everywhere. That was a great step forward.

But you pay a high price for unification. You might reduce the number of natural laws, but it gets more and more difficult for the majority of outsiders to grasp those few fundamental laws.

That's sad but true—though the general public in the seventeenth century had its problems with Newton too. Now every high school graduate understands the laws he discovered. Voltaire made them accessible with his commentaries.

Voltaire's lover, Émilie du Châtelet, deserves the lion's share of credit for that. She wrote most of the commentaries. But anyway . . . does it bother you when, say, at a party no one understands your work?

Yes. Sometimes it's very frustrating. For example, I once testified before Congress on a particle accelerator project here in Texas. And a congressman asked me: "What do you need an accelerator for? Just do it with a big computer." He had no idea that a computer can't calculate anything about phenomena no one knows. With the accelerator we wanted to discover new laws of nature. Even worse are all the people who admire you because you're doing something they don't understand. Many people also believe that inaccessible poetry must be profound. We shouldn't be proud that physics is hard to understand—but do something about it.

What?

Young people are offered only one path to learn physics. For generations they've had to go through the same curriculum. First they're taught mechanics, then something about heat and electricity, then atomic physics, and so on. And everything is packaged in the language of mathematics. That's all well and good for people who want to become physicists and take pleasure in calculating. But for most people, that's not the case. A much more direct way is to tell the students stories that help them comprehend the discoveries of great physicists. With that principle in mind I once wrote a whole book on the physics of the twentieth century. I wanted to revolutionize the way science is taught in school. I thought the book worked well, but it didn't change anything.

Nothing ultimately came of the particle accelerator project in Texas either. An accelerator would have been necessary to produce the very high energies—as were present in the first moments after the big bang—at which the fundamental forces unify.

Unfortunately, I wasn't very successful in political matters.

Still, the standard model has passed with flying colors all tests performed to date using the weaker accelerators of the Geneva center for nuclear research. In 2008 a machine went into operation there, the Large Hadron Collider (LHC), which is supposed to reach similar energies as the one you wanted in Texas. Unfortunately, there was immediately a terrible explosion in the accelerator's underground tunnel and the LHC had to be repaired for over a year. Now the experiments are far behind schedule. Are you disappointed about that?

Moderately. We've waited for this machine for so long that the delay due to the accident isn't such a big deal.

What discoveries are you hoping for?

If anyone knew what would happen, we could spare ourselves the experiments. Most of us expect that a new elementary particle called the Higgs boson will be detected.*

Which is supposed to be responsible for the fact that all things have mass.

But if that's all we find, it would be very disappointing. Personally, I would like to gain insight into dark matter. . . .

A mysterious substance of which there must be five times more in the universe than of all known manifestations of matter.

Up to now we don't have the slightest idea what it consists of. It would also be wonderful to be able to find evidence for a framework

---

* Indeed, the Higgs boson was detected in July 2012 at CERN, the European Organization for Nuclear Research in Geneva. But for now, the full significance of this discovery remains elusive. In particular, physicists remain far from having solved the mystery of dark matter.

called supersymmetry in the realm of the known elementary particles. We've been talking about that for more than thirty years. It would be very exciting to actually see supersymmetry.

The standard model you helped establish back in your red Chevrolet Camaro is not complete. For one thing, all attempts to incorporate gravity have so far failed. For another, it contains numerical values that don't follow from the theory itself but have to be integrated as measurement results. Therefore, physicists have been trying for more than thirty years to find natural laws beyond the standard model. So far they have not achieved that goal.

The standard model is very successful. But take a look at this. [He pulls a red booklet out of his desk drawer.] This contains everything we know today about elementary particles: tables, columns of figures. If this were our theory, how ugly it would be! That's certainly not what we want. Our sense of beauty is part of our job description.

So what would you call a beautiful theory of nature?

One in which the connections arise inevitably. Everything fits together, and if you try to change even a tiny part, the whole edifice collapses. Such theories exist: Just think of quantum mechanics, which describes the dynamics of atoms and elementary particles. The standard model doesn't have that coherence.

Why should nature actually be constituted in such a way that people find its laws inevitable and beautiful?

If I only knew. It's possible it's not constituted that way and we're indulging in wishful thinking. On the other hand, the search for beauty has already brought us far: In the course of time scientists have discovered some very appealing laws of nature. And if we don't keep searching, we certainly won't find any more. I for one find this uncertain goal so attractive that it's worth spending my life pursuing it.

Even if you don't live to see the goal come within reach?

I just want the progress of science to continue—and that can take a really long time. The Greek philosopher Democritus, for example, speculated about atoms in 400 BC. But only 2,300 years later, around 1900, could we be certain of their existence. Democritus's problem was scale. An atom is one hundred thousand times smaller than anything people could see with the naked eye. But today, to make a truly significant advance in particle physics, we have an even longer way to go. To achieve a real unification we would have to explore nature at energies not one hundred thousand times but 10^{17} times greater than what we are currently capable of.

That's an impressive number—a one with seventeen zeroes. But it might not be scale alone that prevents progress in fundamental physics. Your own theory could stand in the way of the search—the standard model was simply too successful. Have you ever regretted your insight in the streets of Boston?

No. Ideas build on each other. The standard model was a step that was necessary before we can take the next one. If I regret a discovery of physics, then it's nuclear fission—for completely different reasons.

What you're striving for is a law that would describe the whole universe. Many people call that a "theory of everything."

I don't like that phrase. It implies that we would understand everything when we've reached that goal. But that won't be the case. Think of phenomena like consciousness or even just turbulence in liquids and gases. We already know the physical and chemical laws underlying them today. And yet we're far from having understood our consciousness or the weather. That's why I prefer to call the goal of our search a "final theory."

Even many physicists have doubts about the value of such a final theory, which wouldn't make us any wiser about some of the most interesting things.

That's unwarranted for two reasons. First of all, our theory will be more fundamental than any other insofar as it applies everywhere. All other descriptions of the world are limited in some way: The laws of hydrodynamics, for example, have significance only where there's a gas or a liquid. But our final theory will apply without limit everywhere in the universe.

Secondly, our theory will be the end of all explanations. In our world we deal with accidents and principles. Accidents can't be explained. It's pointless to ask why a comet hit the earth sixty-five million years ago and wiped out the dinosaurs. It's another thing to attempt to find out something about the rules of heredity among the dinosaurs and all other living things. Those involve underlying principles—to be precise, the principles of biochemistry. And the biochemical laws can be explained in turn by atomic physics. Then comes particle physics, and so on. Ultimately, it boils down to the final theory. That's where all "why" questions end.

Are you familiar with Russian matryoshka dolls?

You mean those wooden nested dolls?

Exactly. The outer doll can be taken apart to reveal a smaller doll inside, which can be taken apart as well—and so on, until you eventually find the smallest, indivisible doll. Your dream of a final theory reminds me of that. But who's to say that the search will ever come to an end? There might, as it were, be an even smaller doll at every level.

That's possible. On the other hand, we see that the explanations at each step we have so far climbed have become more and more encompassing. I find that encouraging. Another question is whether our brains are powerful enough to even understand these increasingly comprehensive laws. In the end, dogs can't be trained to solve the Schrödinger equation.

I wonder to what extent the longing for an all-encompassing law of the universe is our cultural heritage. Jews, Christians, and Muslims believe in the one almighty God. Sometimes the search for a final theory seems to me like monotheism in a new, more secular form.

An interesting idea—after all, science as we know it was born in Europe. But I would turn your argument around: The desire for one God and for a theory of the whole cosmos might have the same cause. Monotheism developed because people found polytheism too complicated. And just as it's less satisfying to pin storms on Zeus, plagues on Apollo, and the crop yield on Demeter, we physicists would rather have a unified explanation of the world than the complex standard model.

Do you mean that the desire for a single reason for everything is inherent to human nature?

We have that trait—though we also have the exact opposite longing. When you go to an opera, you're not looking for simple explanations, but want to experience on the stage the whole diversity and complexity of life.

Different people have different aesthetic needs.

No, it's more that we all bear conflicting needs within us. We want both, simplicity and abundance.

You once described your first encounter with the symbols of higher mathematics as a vision of magical signs.

I went to school in the Bronx. One day I happened to come across a book on the theory of heat in the library and in it I saw this sign: $\oint$

It stands for an operation of infinitesimal calculus, an integral over a closed curve. But I didn't know that. I sensed only that this symbol that was obscure to me must be very powerful—similar to how Faust might have felt at the beginning of Goethe's drama when he stumbles on the pentagram. When I learned that this sign allows heat to be described in precise mathematical terms, I found it even more exciting: Whoever understands such symbols, I thought, could dominate nature.

You dreamed of power.

Not in the sense that I wanted to do something with my knowledge. I wanted to dominate nature intellectually, penetrate its secrets. So I began to teach myself higher mathematics. I felt like a sorcerer's apprentice as my understanding of the symbols improved over time.

In the six decades since, you've achieved more than perhaps any other physicist. Have you often had to struggle with envy?

Not very often. When physicists are interested in what others are doing, it's usually to learn from them. There's not much room for envy—unlike, say, among writers or in the art world. Maybe it has to do with the fact that in art there are not really objective criteria for who deserves fame. So you can easily feel cheated. In theoretical physics, however, there can be little doubt about the value of a particular achievement. When I was still young myself, my colleagues Tsung Dao Lee and Chen Ning Yang from China discovered an astonishing phenomenon in particle physics that became known as "parity violation" and immediately earned them both the Nobel Prize—Lee was not yet thirty years old. I was annoyed because it would have been just as possible for me to make that discovery. But I didn't feel envy.

You simply didn't think of it: You had only yourself to blame.

Exactly. Then, on the day I found out about my Nobel Prize, my wife made a remark that became very important to me. She said, "And now you have to write some unimportant papers."

Because she was afraid that you would lose touch otherwise?

Yes. She wanted to protect me from the effects of excessive admiration—and from getting bogged down in unsolvable questions.

In other words, she was trying to help you avoid the fate of a scientist like Einstein, who spent his late years with a fruitless search for the theory of everything.

Other very famous physicists of the twentieth century became interested in telepathy or Jungian psychology. Compared with Einstein's failure, those scientists' pursuits led them even further astray. The only way to really contribute something to a field is to see it with the eyes of a beginner. That was one reason I later turned my back on particle physics and started afresh, in a way. I wrote papers on cosmology that interested almost no one besides a few experts. But only in that way could I learn new things about nature.

What are your feelings today when you think about nature?

A sense of beauty, of wonder and mystery. However far we come in the search for a final theory, we'll never know why the laws of nature are the way they are. A mystery will always remain.

Many people, including some of your colleagues, call it God.

Not me.

Why not?

Out of respect for our history. The word "God" has had a fairly clear-cut meaning for centuries in the West: It has meant a being of some sort, a creator concerned with questions of good and evil. I don't believe in such a God. When Einstein calls a cosmic spirit of beauty and harmony "God," he is lending the term an entirely new meaning. He seems to me to be doing violence to a well-established word. Ultimately, thinking about nature doesn't fill me with anything even close to the emotions I would have toward a personal God. The laws of nature are impersonal; they're not interested in us. How could I have warm feelings for them as I do for another human being or even for my Siamese cat?

Your book *The First Three Minutes*, in which you explain the standard model, ends with a famous sentence: "The more the universe seems comprehensible, the more it also seems pointless." What did you mean by that?

That we find nothing that gives our lives an objective meaning. There's nothing in the laws of nature to suggest that we have a particular place in the universe. That doesn't mean I find my life pointless. We can love each other and try to understand the world. But we have to give our lives that meaning ourselves. Perhaps you remember that another sentence follows the one you mentioned: "The effort to understand the universe is one of the very few things that lifts human life a little above the level of farce and gives it some of the grace of tragedy."

Why tragedy?

As compared to what was previously believed. Human beings regarded themselves as characters in a cosmic drama: We were created, we have sinned, we will be saved—a grand story. Now we realize that we're more like actors standing around on a stage without direction and we have no choice but to improvise a little drama here, a little comedy there. I experience that as a loss.

One could also view it as a gain in freedom and be happy about that.

If you can do that, my congratulations. I would like to be able to, but I can't. I feel a certain nostalgia for a bygone age of belief. I find myself attracted to religion. And my aversion to religion stems from the fact that I feel a longing for something I know isn't true.

Have you always felt that way?

My parents weren't especially observant Jews. Until I was about twelve, I believed that there must be something like a God, even if I didn't adhere to any particular faith. Then those ideas suddenly seemed silly to me, and that was the end of that.

You've spoken out very sharply: "Anything that we scientists can do to weaken the hold of religion should be done and may in fact in the end be our greatest contribution to civilization." What makes you so angry?

History. I see the decline of science, which had been so productive in Greek antiquity, largely as a result of strengthening Christianity. The Byzantine emperor Justinian I closed the Platonic Academy, because the study of nature was regarded as the surest symptom of an unbelieving mind. In other respects, too, religion seems to me to do more harm than good. Just think of all the religious wars— to this day. Currently many people are afraid of militant Islam; in the sixteenth and seventeenth century Christianity was horribly intolerant. But with age my tirades against religion have mellowed. I never had the feeling that they convinced anyone.

Religions can mellow too. Christianity today is no longer particularly intolerant.

True, it changed in the eighteenth century. . . .

Under the influence of the Enlightenment and science.

Which is why I think that one of our most important tasks consists in weakening religious certainties.

The state of Texas, where you live, is not exactly known for its religious openness. Do you ever feel somewhat lonely with your views?

Not at all. A surprising number of people feel the same way I do. They keep telling me how glad they were to read my essays. And the others are often much less religious than it seems. They don't try to convert you, because they themselves are uncertain about their beliefs.

Maybe they just have good manners.

It's not only that. I once had to present my opinion before the Texas Board of Education on how Darwin's theory of evolution should be dealt with in the classrooms of our state. I recommended disregarding the religious implications and just teaching whatever is good science. We won that debate—simply because the government did not want to appear backward. Serious believers would have decided differently.

Scientific research and belief can coexist peacefully, if they confine themselves to separate spheres. Science is concerned with the measurable and demonstrable, religion with values and visions that elude empirical verification.

Certainly. But that means a tremendous retreat for religion, because it once wanted more. Until Charles Darwin published his theory of evolution, there was a "natural theology," which attempted to prove the existence of God from his creations, living things. Darwin put an end to that argument once and for all. Since then, if God can be explained at all, then it's only metaphysically. But to make that retreat is dangerous for any religion, because the question immediately arises: Why should I believe what cannot be verified? It's for good reason that the commission in the Vatican that investigates the miracles ascribed to candidates for sainthood meets under strict secrecy.

But people believe nonetheless. Perhaps we have a natural predisposition to religiosity.

So what? We're also naturally inclined to fill our bellies with sugar and fat, and we struggle against that.

But that takes strength. How do you live your atheism?

Great works of art can console us.

But many of the artworks that move me most intensely draw their power from the religious faith of the artist. Without it Bach's fugues, Giotto and Piero della Francesca's frescoes, or *The Divine Comedy* would never have existed.

I feel the same way: It is a loss. No architecture moves me even close to the way Gothic cathedrals do. Sometimes religion seems to me like a somewhat crazy old aunt who lies, gets up to all sorts of mischief, and might not have much life left ahead of her. But she was once very beautiful—and when she's gone, we'll miss her. Still, we can go on enjoying cathedrals and Gregorian chants without believing. And many of the greatest pieces of literature manage without any religious background; just think of all the works of Shakespeare. And in the end we still have humor. In one of Woody Allen's movies—I don't remember which—the hero goes through profound existential angst. Eventually he ends up in a movie theater where a Marx Brothers film is showing.

That's in *Hannah and Her Sisters*.

The humor on the screen reconciles Woody Allen's protagonist with his life. We can be amused with ourselves—not with a sneering humor but with a kindhearted one. It's the sort of humor we feel when we see a child taking its first steps. We laugh at all the child's arduous efforts, but we do it full of sympathy. And if laughter ever fails us, we can still take a grim satisfaction in the fact that we are able to live without wishful thinking.

Can We Become Immortal?

Molecular biologist

Elizabeth Blackburn

on aging

◆

AGING SEEMS TO BE ONE OF the unpleasant facts of life—though the sixteenth-century French essayist Michel de Montaigne didn't think so. He wrote, "To die of old age is a death rare, extraordinary, and singular," granted to only a lucky few in Montaigne's violent and plague-stricken time.

Today scientists question whether our physical and mental decline is really inevitable. Elizabeth Blackburn is among the pioneers of that research. Born in a small, remote city in Tasmania in 1948, the second of seven children, she studied biochemistry at Cambridge. She went on to investigate the genetic mechanisms of aging. For that work, she received the Nobel Prize in 2009.

In her lab at the University of California, San Francisco, she talks about her discoveries as enthusiastically as if she had just made them. At those moments, you sense behind Blackburn's friendliness and sense of humor a formidable tenacity. Her gently uncompromising character has not only advanced her career; it also led to her dismissal from President Bush's council on bioethics in 2004.

◆

Dr. Blackburn, you've devoted decades of your life to ciliates. What's so fascinating about single-celled organisms?

They're marvelous creatures. They can reproduce asexually by simply doubling themselves. Yet they have seven sexes that mate in pairs. And occasionally even three ciliates join to breed. That makes you wonder why we're content to be women or men—especially since ciliates, regardless of sex, have a choice of seven different types of mating. That's wild! How can anyone not love those organisms?

Did you suspect that with your investigations of ciliates you were close to solving the mystery of human aging?

No. We wanted to study fundamental questions of molecular genetics. That was exciting enough: In 1975 my later husband and I were among the first people who could read genetic information at all. Ciliates were well suited for those experiments. As my lab made progress over the years, I started to think that we were getting at the heart of biology. But I never had the goal of curing human aging.

Reading the newspaper articles on your 2009 Nobel Prize, one might think that you had managed to do just that. Did you find all the hype excessive?

Not at all. For a long time, I myself couldn't believe that the discoveries we made about ciliates could be applied to humans. But we've since found unequivocal evidence.

Ciliates are immortal.

That's part of the beauty of their biology. These single-celled organisms can divide endlessly, perpetually beginning a new life. We wondered how they do that. The problem is that a bit of the chromosomes, which contain the genetic information in the cell, is lost with each division.

Eventually they become too short, and the organism can no longer function.

That's just what the ciliates prevent with an extremely well-functioning repair mechanism. Carol Greider, my graduate student at the time, found evidence for that on Christmas Day in 1984: There's a substance in the nucleus of the ciliates that is able to perpetually rebuild the ends of the chromosomes.

An elixir of immortality for cells.

We called it telomerase. It helps form a sort of protective sheath on the chromosome—the telomere, to which ciliates owe their endless life.

Our bodies can regenerate as well. Our organs too rejuvenate themselves through cell division; their cells, in a sense, produce their own successors. Only that doesn't happen as often as we'd like.

Exactly. As we age, more and more cells die off without replacement. As a result, our bodily functions deteriorate. But humans have telomerase too. Ten or so years ago, scientists found that in families with members who don't produce enough telomerase due to a genetic disease, those members suffer unusually early from age-related afflictions. That proved that telomerase delays aging for us as well.

Do those poor people get premature dementia?

Unfortunately, they don't have enough time for that. They die beforehand of cancers and all sorts of infections—as if their immune system simply runs out of steam. It seems to be related to the fact that their telomeres get too short. Since that discovery we've witnessed a tsunami of insights on the connection between aging, diseases, and telomere length.

It's as if inside each cell were something like a thread of life. In Greek mythology this dictates not only the length but also the quality of our lives.

A nice image. But the development doesn't always go only in the direction of decline; occasionally telomerase causes the telomeres to grow again.

What determines how well our cells regenerate?

The circumstances of life play an important role—especially chronic stress. In collaboration with psychologists, we studied mothers of disabled children. Often here in the United States they don't get much support, and they're under enormous stress. And the more years they took care of their children, the shorter their telomeres tended to be. We found a similar situation among people who had suffered trauma as children, like the death of a parent or even sexual abuse. The greater the number of terrible experiences they had to cope with, the more their telomeres had shortened on average.

As if each blow of fate cut off some of the thread of life.

Stress early in life seems to leave particularly deep traces in the nucleus. These results make one thing quite clear: how critical it is to protect our children. There are people, however, who can get over even great hardship amazingly well.

Apparently, how long we live is also hereditary.

Yes. Fabulous evidence of that is the *Gotha*, the German almanac of the nobility. In it, the life spans of around five thousand daughters from all over Europe are recorded; those of the sons don't tell us much because too many of them died at war. The women, on the other hand, almost always led comfortable lives, as long as they survived childbirth and childhood infectious diseases. If you compare the age they reached with that of their parents, you make an amazing discovery: Up to about seventy-five years, the one has little to do with the other. Whoever dies up to that point fell victim by chance to an illness or a misfortune. But whoever makes it beyond her seventy-fifth birthday has her genes to thank: Typically, those long-lived nobles also had particularly long-lived ancestors.

All of us are increasingly in the situation of those noble daughters. Thanks to good hygiene and medical care the likes of which not even queens could have hoped for in earlier eras, most people could easily make it to seventy-five. Do genes then set a natural limit to our lives?

We might be the first generation that can find out—because we live in an environment more protected than ever before. It's not what we were selected for in our evolution. Even though lots of people still die of cardiovascular diseases, the number is decreasing; heart attacks can be avoided through a healthier lifestyle. Whether that also goes for the other big killer, cancer, is an open question. How far human life expectancy can be increased is a huge experiment. And all of us are the lab rats.

What's your hypothesis?

To live to 120 is clearly permitted by the current gene pool of our species. The oldest person up to now was Jeanne Calment from the South of France. She learned fencing at eighty-five, rode a bike at one hundred, and died in 1997 at the age of 122. As is true of most people over a hundred years old, her family members lived to be very old as well. And she enjoyed excellent health all her life— even though she smoked like a chimney!

What did she die of?

It's unknown. It's possible that her death had no particular cause, as is the case with so many really old people: Eventually the whole system simply becomes unstable. Then all it takes is a fall or pneumonia and you die. On the death certificate they then write "heart failure." That way, as a doctor, you're always right.

Usually, medical science assumes that diseases and not old age lead to death. You make the opposite claim.

What does the word "disease" actually mean? It can mean different things: On the one hand, there are ailments with a clear cause. We're infected with some bacterium or virus, and the symptoms set in. That's where medicine has had its great successes. On the other hand, there are cardiovascular diseases, cancer, and adult-onset diabetes, which arise from the organism itself and progress over a long period of time. Currently, doctors can usually only help people live with those afflictions—if they can help at all. Our medical establishment focuses too narrowly on the symptoms. The diabetologist tries to deal with your diabetes, the cardiologist with arteriosclerosis, and so on. But underlying these diseases is a more general process: the failure of the body's own repair mechanisms.

And with your research you hope to find a new approach to those afflictions.

Yes.

Which would be?

The three big killers of the elderly, cancer, cardiovascular diseases, and diabetes, clearly are influenced by the state of our telomeres. To better understand that, we've partnered with human geneticists and a big American health care provider. By scouring the medical histories and habits of one hundred thousand people whose average age is about sixty-five, analyzing their genes, and measuring their telomeres, we hope to find out how combinations of environment, lifestyle, and genetic makeup impact health.

People don't object to their health care provider organization investigating them so thoroughly?

They all signed consents. And we noticed that during some preliminary studies, people came storming in our doors in their eagerness to participate. Some urged us that they were of particular scientific interest because of their CV or daily yoga practice. All of them wanted to find out about their telomeres.

Do you enjoy the role of modern palm reader, who predicts when people will die?

But no one will learn that from me! Telomere length by itself predicts a particular life expectancy only in statistical terms, just as someone with high cholesterol will with greater probability—but far from absolute certainty—suffer a heart attack. It is the combination of many factors that matters. Unfortunately, many people have a hard time understanding statistics.

Because they are understandably not interested in how many out of a hundred patients with a particular telomere length will still be alive in five years. They want to know their own fate.

Especially as it's so seductively vivid when you imagine that telomere length directly corresponds to biological age!

You've cofounded a company with the aim of offering such tests to everyone. Why?

Because there's a demand for it. Our university lab could no longer manage the volume of inquiries; that's how it all began. And someone has to be the first to go public. Better we do it right than others do it wrong. That's why you won't be able to send in your sample yourself, but only via your doctor.

What good will this do me?

You get information about your body.

But there's not much I can do with that information. You've only just begun to investigate what exactly the state of our telomeres means for our health.

And we don't claim to know the answers. Everyone who participates is to be informed that their data serves the ongoing research.

The test subjects risk results that can be extremely depressing. Would you want to know that there's a 90 percent probability you will die in the next five years?

We conducted a preliminary study on that too: No one seemed particularly worried about their test results. Experience with other genetic tests that make only statistical predictions also shows that people are able to deal quite well with the results. If you have short telomeres, that's just a warning sign to take a closer look—like the check engine light on a dashboard.

Can we do anything as adults to reverse the degradation of our telomeres—or at least stop it?

Our studies on that are still in their infancy. One thing, at least, is clear: People who exercise more and sleep better have longer telomeres.

Do you know the length of your telomeres?

Yes. I'm not worried. I'm okay.

Have the results changed your lifestyle?

No. But since the results of research on telomeres in people started coming in, I've tried to exercise thirty minutes a day. That's the only magic bullet against physical decline that I accept. The evidence for it is compelling.

Just exercise? That makes you a minimalist. Even leaving aside the billion-dollar nutritional supplement industry, scientists advocate a whole range of prescriptions to combat aging: less sugar, more red wine, vitamin E, green tea....

Everyone takes what seems helpful to them. But unfortunately, no one can ever prove the effects. No one knows how polyphenols, the substances in red wine and green tea that are supposedly so beneficial, really work—or whether they work at all. High amounts of vitamin E can cause cancer. I'm sure it's sensible to avoid eating too much sugar. But quality of life is important to me, too.

You're sixty-three. Do the signs of aging bother you?

I find my age excellent. When I was your age . . .

. . . forty-six . . .

. . . I was the mother of a young child. At the same time, my research made extremely heavy demands on my time and energy. I was horribly stressed. Besides, I now have a broader perspective on the world. I wouldn't want to trade places—even if I was a better skier back then.

Many people find it humiliating that their skills diminish in old age and their looks aren't what they used to be. Women in particular suffer as a result of that.

But it doesn't have to be that way. Women can bloom again when the kids are out of the house. For that they need support, however. Unfortunately, it's still common in our society that older women aren't valued.

You can hardly complain about that—with all the honors you've attained.

In the United States, Nobel laureates are no great rarity—but female Nobel laureates are. When I appear in public, lots of people get really excited and want their children or grandchildren to see me. The mere fact that I exist sends a message that women can achieve things of this significance. So I realized at some point that I don't have to do anything else to be useful. It's enough to stay alive.

Would you like to live to be 120 or even 200?

Oh, yes! But you have to think about what age you want to stretch out. I would gladly turn down the prospect of extending the stage between eighty and ninety. Most people would probably like best to increase the time between twenty and thirty.

Not me. Too much lovesickness.

On the other hand, in those miserable years, we're at the height of our mental capacity. If I really think about it, I'd like to start over repeatedly with a twenty-year-old brain. First I would spend twenty-five years redoing what I did. Then I'd try to master mathematics and go into cosmology. In cosmology there are such exciting open questions that I sometimes think: Why are you wasting your time with biology? I would play piano more often. And go skiing a lot more.

> Leon Kass, the chair of the council on bioethics, of which you were a member, found such dreams disturbing. It is precisely transience, he argued, that has brought out the best in human beings: engagement, seriousness, ties between parents and children. What did you reply to him?

That short life is damned inconvenient. My colleague assumes that, with the prospect of more years, people would become lazy. But can he prove that? I certainly don't know what in my vision of three successive careers suggests a lack of seriousness. However, I'm not saying that my model works for everyone. Some people would be worried that they'd have to be married to the same person for 180 years!

> But the question is what the point of aging is in the first place. Turtles, for example, are spared that fate.

In what way?

> Not even an expert can tell the difference between the organs of a young turtle and those of a hundred-year-old turtle.

Apparently, those amphibians have extremely efficient repair mechanisms. In evolution, different reproductive strategies have prevailed. An animal can either reproduce mainly in its younger years, as we do, in which case a longer life offers no biological advantage, or it can produce offspring until it dies, as the turtle does, in which case each year is a gain. However, it costs the organism a lot of energy to constantly stave off decline.

> Food might have been scarce for our ancestors, but it's not for us. Would it be conceivable to improve human metabolism to the point where we wouldn't age anymore either?

In theory, yes. We could become immortal. The only question is whether our cellular machinery would be adequate. It's possible that the system we're born with would eventually reach a point where nothing else could be tweaked.

Where might that be?

We don't know. I'm on the advisory board for an initiative called Tara Oceans, which is attempting to study all the life in the ten meters under the ocean's surface. People find the most amazing creatures there—such as copepods with incredibly well-functioning repair systems. Even though they're multicellular organisms, some of them might even be immortal.

It's not a law of nature that higher life moves inexorably toward death.

No, it's not.

Are you afraid of death?

Not anymore. My son is grown up. I'd feel sad for my husband and him if I passed away. But a lot of good things have happened in my life. Why should I fear death?

Is Luxury Immoral?

Philosopher

Peter Singer

on ethics

◆

OPINIONS ARE DIVIDED ON PETER SINGER. His supporters regard him as one of the most significant moral philosophers of our time and praise his bold thinking. Among them are many of his colleagues; vegetarians, who view him as a pioneer of the animal welfare movement; and Bill Gates, who sympathizes with Singer's unconditional commitment to the poorest people in the world. His opponents, on the other hand, fault Singer for failing to recognize the sanctity of human life. His fiercest critics even claim that Singer's ideas come dangerously close to the ideology of National Socialism—despite the fact that Singer comes from a long line of Czech rabbis. His parents fled from the Nazis to Melbourne, Australia, where he was born in 1946 and, after completing his studies, became professor of philosophy. He's now a laureate professor at the University of Melbourne and a professor of bioethics at Princeton.

Most of the conversations in this collection were conducted face to face. Singer, however, was reluctant to meet in person—due to moral considerations about the environmental impact of a transatlantic flight. Couldn't we talk via Skype instead? A first.

◆

Professor Singer, what is the argument against my flying from Berlin to New York to talk to you?

I think it wouldn't be morally acceptable; we should do what we can to reduce our carbon footprints.

A one-way transatlantic flight generates a little more than four tons of CO_2 per passenger—in business class it's more than six tons.

Since there is an alternative—there are other ways of doing this interview, as we are finding—unless there was a very pressing reason why you would have to do it in person, I think it would be wrong to travel when you don't need to. I'm skeptical about whether a face-to-face interview as compared to a Skype interview would make a difference in the quality of your work.

How do you usually handle travel? After all, you're a professor in Princeton as well as in Australia.

I'm not a saint. Lately, when I speak at conferences and things like that, some organizers pay my fare, and if it's business class, I accept it. Especially on long flights, the seats are much more comfortable. I find the economy seats pretty cramped, and I don't arrive in very good shape because I can't really sleep in them. I think it's ethically better to fly economy class, but there are cases where self-interest and morality clash, and people will often choose self-interest in those cases—myself included. But let's say I was flying and I needed to get to a talk within a day or two, and this was an important talk. If I might be better able to influence people who could make a difference in the world if I arrived fresh and well rested rather than tired, then maybe business class could be justified. But if it's only for my personal comfort, it's probably not possible to really defend it.

What exactly is your understanding of morality?

We behave morally when we improve the lives of all we can affect. We have to take into account not only our contemporaries, but also those who haven't even been born yet—to the extent that we can predict the consequences of our actions.

The total benefit must outweigh the harm.

Right. So I find it hard to see how that calculation is supposed to work out when someone, say, flies from New York to Thailand for three or four days just to take a holiday there. I think, in the world we live in today, that is really quite seriously wrong.

You're a utilitarian—that is, for you there's no right or wrong from the outset. You judge decisions by their consequences.

I'm not sure what "from the outset" could mean here. To tell right from wrong, it is not enough to just give a rule—you have to then justify that rule. Consider, for example, the famous case, which Kant himself discussed, about somebody who comes to your door wanting to murder an innocent person, and that person is in fact hiding in your house. Kant said you should not tell a lie. But I think most people would think the right thing to do under those circumstances is to lie. So the rule "Do not tell a lie" is not actually a clear rule, because, in the way we really think about and accept it, it's "Do not tell a lie unless . . ."—and then the "unless" has to be spelled out.

But when deciding if an action is morally acceptable or not, in general it's very hard to be aware of and to calculate all the consequences. And exactly who is to judge the possible consequences of actions?

Well, I think you are right. That's why many utilitarians—and I would include myself here—would say that, because we are likely to be biased judges, we should accept the general rules that prevail, unless there are other clues or pressing evidence.

As a utilitarian, you must subscribe to the argument that torture can be moral—for example, if it's the only way to prevent a crime.

I would say such practices were abused under the Bush administration. Still, if we had really strong evidence that it was the only way to save thousands of lives, it would be hard, I think, for a utilitarian to say, "No, that's still wrong."

But isn't the problem that we can't determine in advance when a suspicion is very well founded, or whether the cruelty against the suspect will really achieve what's intended? That's why I think it's right to forbid methods like torture no matter what the circumstances. Several years ago, the Frankfurt police arrested the kidnapper of Jakob von Metzler, the son of a banker. But there was no trace of the victim. The deputy police chief gave orders to threaten the prisoner with torture in order to make him talk and to save the boy. In the end, the kidnapper led the police to the child's dead body. The police chief was found guilty of coercion.

Any civilized society ought to prohibit torture, because if it is permitted, it will be abused. And even though the police chief might have been right to do what he did, since there are rules against this, you have to be prepared to accept the consequences of breaking them.

Now you're making the case for a double standard. Someone who tortures behind closed doors is not necessarily doing wrong—but if the matter becomes known, society has to punish him.

As a utilitarian, you can't always be against double standards. On the one hand, we should have a public rule that says torture is forbidden, and on the other hand, there are some rare cases in which torture may be justifiable.

For a long time, your colleagues wrote only abstractly about ethics. You, on the other hand, deal with practical questions of moral philosophy: vegetarianism, poverty, doctors' decisions about life and death. What made you take that step?

I don't deserve all the credit for reconciling ethics with praxis. But I contributed to that development. And it's not that it never happened before, either. I mean, there is a lot of philosophy that you can read that is very practical. Thomas Aquinas gave decision-making advice, David Hume wrote about suicide, Kant discussed the practical implications of all sorts of things, such as sexual morality and how we should treat animals. All that became unfashionable in the middle third of the twentieth century. Suddenly there was a view that philosophers shouldn't try to answer moral questions at all anymore; that was the role of the preacher. I think that is a mistake.

Do you want to change the real world?

Yes, I do. I think that began when, as a student in Melbourne, I was very preoccupied with politics. Australia was fighting in Vietnam at the time, and there was conscription. And so I was learning philosophy while at the same time being confronted with questions like, "When is war justified? Are citizens obligated to obey the state when they think the state is acting wrongly?"

You then became known for a 1972 essay on poverty. You argued in it that it's immoral to live in luxury while people elsewhere lack the bare necessities of life. To this day, it's your most-cited work.

After an uprising in what is now Bangladesh, nine million people had fled to India, which, being a poor country, was overwhelmed by the influx. At the time, I was living in England on a student scholarship. But I found it intolerable to spend money on things I didn't absolutely need while millions of refugees didn't have shelter or clean water and many of them were starving. This led me to think that this is an important practical issue: Why would we go about our comfortable lives and give nothing or put some trivial amount into a collection box, when others are starving or lacking shelter?

What did you do?

My wife and I decided that we would give 10 percent of our income to organizations helping the refugees. Since then we have gradually increased that percentage for the benefit of the world's poorest people. And I tried to explain in my essay that there's something fundamentally wrong with the way we're living.

> In your book *The Life You Can Save*, you compare us residents of the rich North with someone who witnesses a little girl drowning in a pond but doesn't rush into the water because he doesn't want to ruin his shoes. I assume that utilitarianism says that we're morally responsible not only for what we do, but also for what we fail to do.

Exactly. If you'd rather drive a Mercedes than a cheaper vehicle, there are implications: You bear some responsibility for having spent money you could have used to save lives.

> According to an estimate by the United Nations, an additional thirteen billion dollars a year would be enough to establish basic health care for everyone in the world. That corresponds pretty closely to the amount we Europeans spend yearly on ice cream. I find figures like that encouraging, because they show how much we could accomplish—and without even giving up that much. At the same time, they're disturbing. . . .

Because they show that our declared belief that every life is of equal value is only a theoretical belief. It's not a practical matter—it doesn't influence our actions very much.

> Our psychology stands in our way.

Yes. We have two ways of getting people to act. Reasoned discussion using words and arguments is the logical, analytic process, but we also have an emotional process. It could be that instead of this long discussion, I could have shown you a very thin child and said, "You can help this child if you give the child food."

We see more than enough photos of poor children we can help with a donation. They stare at us from billboards everywhere. But we still don't donate much.

Psychologists call it the diffusion of responsibility. Because we know that every other passerby could help, and that some are even better off than us, we feel it's not up to us. In the end no one feels responsible. We hide in the crowd. In New York there was the famous murder of Kitty Genovese in the 1960s. The young woman died even though a whole apartment building heard her screams. But no one called the police because everyone assumed someone else would do it.

And even when we do help, our natural responses often lead us astray. The results of one study you highlight in *The Life You Can Save* suggest that people would rather give money for medical treatment that would save the life of a single child than for treatment costing the same amount that would save eight children.

You know that there is one child in need and that you can save this child, and so you feel good that you can make a difference and solve the problem. But if you tell people that for the same amount of money they could save a larger number of children but that there are more children in need—that there are one hundred people in need and you could save twenty of them—fewer people would save twenty than would save the one. Any economist would say that's crazy. You obviously get better value for your money if you save twenty children than if you save one. But at twenty out of one hundred, a lot of people just see the fact that they can't really make a difference, so it's not worth it to bother.

Apparently our intuition isn't much help in eliminating as much suffering as possible. Maybe that's partly because our moral impulses developed under different circumstances from those of the present. Our brains were programmed when our ancestors lived in small communities. Something like adversity we only know about but don't experience directly might not even have existed back then. Our preference for problems we can solve completely might also be innate. We shrink from the rest.

Certainly, our moral views are not well adapted to the world in which live. That's why it's so hard for us to understand emotionally that a flight to New York in business class is morally just as wrong as poisoning the fields of someone who needed the food to live. Greenhouse gases might lead to long droughts in Africa that will make it impossible for subsistence farmers. According to the World Health Organization, 140,000 people a year die from the spread of infectious diseases due to climate change, and that's only the beginning of what's likely to happen over the next century or two if we don't control greenhouse gases.

Do you think people are able to translate such arguments about practical ethics into better decisions?

I think people can be influenced by the things that I write. On the website I set up, www.thelifeyoucansave.org, the last time I looked it was something like 17,900 people who have pledged to contribute according to the levels that I suggest in the book, and we know that as a result, millions of dollars have gone to highly effective charities.

Even as you've attempted to persuade the public with your arguments, you haven't made it easy for everyone to accept your ethical stances. Many people associate your name with views they find repugnant—specifically, your opinions regarding newborns and people with disabilities. At your appearances in Germany there were protests, and when Princeton appointed you professor, outraged donors of the university withheld their financial backing. Wouldn't it have been better for you to have kept to yourself your opinions on bioethics—even if only in the interest of the global poor?

That's an interesting question, and the question could be looked at in different ways. If the consequence of my views on bioethics is that it became harder for people to accept my views about giving to the global poor or something of that sort, that's a question you can only ask with the benefit of hindsight. Another question is whether I could have anticipated when I wrote about my views on bioethics that this would happen—that's a different question. To go back to the first question, on the one hand, no doubt it's true that there are a lot of people for whom my name has negative associations because of my views on bioethics. On the other hand, probably a lot more people have heard of me at all because of these views. In that respect, the controversy helped me get my views on all kinds of topics across to a wider audience. As for the second question, at the time when I first put them forward, it didn't seem to me that my views on bioethics were really likely to be so controversial.

Really? "Very often [killing a disabled infant] is not wrong at all," you've written. I find that sentence shocking.

You've taken it out of context. Actually, for years it didn't raise any controversy. That sentence appeared in my book *Practical Ethics*, which was first published in 1979. There weren't protests about it until 1989, when I gave talks in Germany. At that point, the disability rights movement, which hadn't even existed yet when my book was first published, had begun to pick it up.

With the sentence I just quoted, you concluded a section in which you argued that killing a newborn diagnosed, for example, with hemophilia, could be the right thing to do. Would you still take that position today?

I think it's a widely shared view that a life without hemophilia is better than one with it, and my evidence for this is that when hemophilia is diagnosed in a fetus by genetic testing, the overwhelming majority of women decide to terminate their pregnancy. And while I see the distinction between killing before birth and killing immediately after birth—it's like the difference between withdrawal of life support and active killing—I still think that if people are prepared to do this to a fetus, it isn't so different to do it to a newborn.

Doesn't it make a difference in the amount of physical and emotional suffering for the mother if she terminates her pregnancy after, say, the second or third month rather than terminating it after the seventh or eighth month—or even giving birth to the baby and then having it killed?

I think it's better to have an earlier abortion than to have a later one, but I still think it's not a decisive moral difference. Assuming the mother knows what she's talking about because of some family experience with hemophilia, that she doesn't detect the hemophilia before, that she thinks she doesn't want to raise a child with the disorder, and no adoptive family can be found—I'm not really advocating that any mother do this, but I think it's a defensible decision for her to make. If parents feel that they can't really love and care for this child, I think it's wrong to impose the child on them. Of course, if another family is willing to adopt and love that child, then that is the best outcome. But that is not always the case.

But people are notoriously bad at predicting how they will react emotionally to an unknown situation. For example, everyone knows what it's like to sit in a wheelchair. But if you ask people how they would feel if they were paralyzed from the shoulders down in an accident, many of them answer: "I'd rather be dead." In fact, paraplegics get used to the new circumstances of their lives remarkably quickly. They go through a stage of intense depression, but after less than a year most of them are nearly as satisfied with their lives as they were before the accident. So how can the parents of a newborn know what emotions they will feel toward their child later on?

I would give the parents the information you've just presented: that almost everybody adjusts to their condition and feels their life is worth living.

Why leave it to the parents to decide whether the child lives? What made people in the disability rights movement so angry, if I understood correctly, was that if a certain condition, whether hemophilia or Down syndrome or some other disability, is regarded as a sufficient reason to terminate a life, that suggests that the lives of people who have this disability have less value.

Then they should be angry at anyone who engages in selective termination of pregnancy for just the same reason. And it seems that about 90 percent of the population will do this, if they find themselves in the appropriate circumstances.

Do you think ethics has to be consistent?

My short answer is yes—if your ethics is inconsistent, something is wrong somewhere. You might say therefore that it's inconsistent to have one standard for the eight-month fetus and one for the newborn. To me, it's more a matter of saying, "Here's a problem. What's the best way of responding to that problem?" I discussed the issue of infanticide when doctors from a neonatal intensive care unit approached me. At the time, in the 1970s, a malformation known as spina bifida was relatively common.

Children with this affliction were born with an often untreatably damaged spinal cord and almost always died a slow death, unless they were given aggressive medical treatment, which kept them alive, but left them very severely disabled.

Option one is you treat every baby with all available medical techniques to try to prolong its life and give it as good a life as possible. Since most of these babies were still not going to have lives of what seemed to us to be acceptable quality, the other option was, once you've made the assessment of which are the most severely affected, you decide that they're not going to be actively treated. But in this particular case nontreatment didn't produce a better or humane death; it produced a long, drawn-out death. And so once you've made that decision, difficult though it may be, why not make sure that the baby dies swiftly and humanely?

So, then, can it be acceptable to question the right to life of all newborns because of such borderline cases? Couldn't this be the beginning of a descent down a slippery slope, allowing much more widespread killing?

I don't see the danger of a slippery slope. In the Netherlands euthanasia has been permitted for some time. Yet I don't see any deterioration of respect for other human beings. They're just more honest about what they're doing. I have not seen evidence that it's causing any kind of increase in unwarranted killing. There is such a thing as compassionate killing. While you might worry that it would reduce the taboo on killing other human beings, I would argue that we would extend compassion and concern for preventing suffering.

> But the burden of proof is on you. If you want to abolish a prohibition as fundamental as that, you should be able to prove what the benefit would be. Otherwise, I'd feel better if, as a precaution, we stuck with the rule that you can't kill people.

There are some concrete benefits, I think, such as reduction of pointless suffering. Apart from that, the task of proving potential harm and benefit is impossible. Moral standards can be applied in all sorts of contexts, and we haven't always known the risks of changing these standards. For example, we are now much more permissive sexually than we were a few decades ago. Many people thought that this would be terrible in various ways, generally speaking, but it doesn't seem to have been so terrible, does it?

> In that respect, any ethics is unsatisfactory—because it is based on suppositions, not on established facts.

Yes, and I wish that weren't the case. I would love to have better scientific studies on the consequences of anticipated reforms.

> Why should we behave morally—in the sense of not putting my interests above the interests of everybody else—in the first place?

There's quite strong psychological evidence that people who are living that way are happy, that people who are generous are more satisfied with their lives than those who are not. This is a justification for acting morally that is broadly based on the better aspects of human nature and the prospects for living a life that we regard as worthwhile and meaningful.

> New neurological investigations go even further. They show that the decision to share voluntarily with others can trigger a sort of pleasure

circuit in our heads—the same mechanism that produces our emotions when we enjoy a piece of chocolate or have good sex.

That's really interesting. Some philosophers think that findings like that undermine morality. They think that if you are influenced by the idea that you will be happier by doing the right thing, your actions no longer have true moral worth. I see that as a pernicious misconception of what morality should be about. Enlightening people about what will be in their own best interest—rather than just telling people, "This is your duty and you must do it"—I think that's definitely a part of morality, and the better way to go.

And what do you do with your guilt when you've once again acted contrary to your moral insights—by, say, flying unnecessarily from Germany to New York for a weekend?

Well, you can pass up your next opportunity to go on a trip, give the money you've saved to an aid organization, and see how that feels. Maybe it will actually be more satisfying than the weekend trip to New York, and instead you can have a quiet weekend at home or in your neighborhood where you can catch up with friends and other things, and you will have the fulfillment, the satisfaction of knowing you have contributed something significant to reducing global poverty. The way to have a good character is to develop it gradually, make a habit of it.

Our Well-Being Depends on Our Friends and Their Friends

Physician and social scientist

Nicholas Christakis

on human relationships

◆

NICHOLAS CHRISTAKIS IS AN EXPERT ON human relationships. But instead of exploring them through empathetic conversations, he uses mathematics. The tools of his trade are computers, in which he has saved data—much of it from online networks like Facebook—on tens of thousands to millions of people: their interests and tastes, the state of their health, and their social networks. Such investigations, Christakis asserts, will spark a Copernican revolution: Just as the telescope once opened up unsuspected worlds to astronomers, so too will this research completely change our understanding of human behavior.

As a trained physician specializing in palliative medicine, Christakis cared for dying patients for two decades. More recently, he was a professor of medicine and sociology at Harvard before moving to Yale in 2013. Today he directs Yale's Human Nature Lab and is the codirector of the Yale Institute for Network Science. According to *Time*, he's one of the one hundred most influential people in the world. We met on Crete, where the Greek-American Christakis was visiting his father. For our conversation, one of Christakis's friends invited us to sit on his terrace overlooking the Aegean Sea.

◆

Professor Christakis, have you known our host a long time?

We met a few years ago. At the time, my father had an electrician at his house and told him about his son at Harvard. "That's funny," the man said. "I was just working in the vacation home of another Harvard professor in this village." My colleague and I hadn't known about each other. And it turned out he had grown up in the area! But that's just the way social networks work: via intermediaries. So I got in touch with the professor, Petros Koutrakis, in Boston. Since then we've become good friends—and our children have too.

What is a friend for you?

Someone I spend my free time with or discuss important matters with. That's the official definition at least. For me personally, the emotional connection is more important than joint activities. I also admire a certain kind of dignity in my friends, and it's best when they share my joie de vivre. The fact that relationships for me are based on conversations might be a somewhat feminine trait—and also a Greek one. You put a bunch of Greek men in a room together. . . .

And they'll talk their heads off.

Yes. Also, my best friend is my wife, Erika, by the way.

Reading your work, I couldn't help thinking of the Roman philosopher Cicero: "A friend is like another self." But you actually go beyond that. You claim that our friends actually make us who we are.

You can put it that way, yes. How happy we are, what movies we like, fatigue, backaches, depression, drug use, even when we die— all that has to do with our friends.

Usually our genes, society, even God, are held responsible for those things.

I don't claim that our social environment alone determines our lives, of course. But its influence is far greater than we think.

In 2007 you first caused a stir with your finding that obesity could be contagious.

The response was incredible—and it was interesting how varied it was. The *New York Times* headline was something like: "Are you packing it on? Blame your friends." A British paper saw it the other way around: "Are your friends gaining weight? Perhaps you are to blame." My colleague James Fowler and I even received death threats.

Why did people want to kill you?

We were accused of contributing to discrimination against the overweight. But that really hadn't been our intention.

Your results were provocative. You claim that it's not just because of junk food and lack of exercise that more and more people are gaining weight: You can catch obesity roughly the same way you get the flu. How did you come to that conclusion?

We used data from the famous Framingham Heart Study. In that town outside of Boston, epidemiologists had been keeping track of the state of the citizens' health and the circumstances of their lives since 1948. But we were able to computerize some previously unused paper records and reconstruct people's social interactions, which took two years. So we knew who worked with whom, who was friendly with whom, who was married to whom, and also who suffered from which illnesses. . . .

The subjects of that ongoing study are like contestants on *Big Brother*.

Now we also have data on the children and grandchildren of the original participants. They're scattered throughout the United States, but we know where they are. We had to computerize almost one hundred thousand handwritten addresses. Anyway, we found out that whole groups of friends are overweight. Having an obese friend increases your risk of gaining weight yourself by 57 percent. And that's true even when the friend lives elsewhere and in a totally different milieu. So it couldn't be due to environmental factors alone.

What if they were already overweight before becoming friends? Maybe heavy people simply gravitate toward one another.

We know that most of them were already friends before either of them gained weight. Do people discover together the pleasure in overindulgence? Of course. But you can also be affected by people who are thin and don't eat much themselves but have relationships with overweight people.

How does that happen?

Through changes in attitude toward body size. Why should you restrain yourself if your friend unconsciously signals to you that a few pounds more aren't really that bad? Your friend's own weight is secondary. It's like becoming infected by a germ from a friend who is asymptomatic but is a carrier of the germ he got from someone else. Later we made similar findings about smoking and happiness.

Happiness is contagious?

Exactly. Like overweight people, happy and unhappy people cluster together. But here too their mood depends not only on that of their friends, but also on that of their friends' friends.

That might also explain where the Germans get their perpetual dissatisfaction or the Americans their chronic optimism; both are mysteries for the scientific understanding of happiness. If we Germans

are grouchier, it might be because we're always infecting one another with our bad mood.

Yes, why are the Germans so German? It's not enough that we're raised German or American as children. Every culture is rooted in a social network in which people reinforce one another's attitudes.

As city dwellers we can choose the people we surround ourselves with. If happiness and habits are contagious, it seems tempting to simply change your circle of friends in order to improve your life. How many friendships have you ended for that reason?

None at all. I'm a friendly person. Such drastic measures usually don't work anyway. If you decide to sever your ties to an obese friend, that doesn't help you lose an ounce—because the loss of a relationship also contributes to weight gain. We've found evidence for that. On the other hand, if your friends are armed robbers, it might be good to avoid them.

How did you get interested in searching for the influence of our friends on our lives?

My mother was diagnosed with a terminal illness when I was six years old. That was why I studied medicine, like so many people who grow up with a chronically ill parent. Because I wanted to do something for people who were dying, I specialized in palliative care. That experience confronted me with the widower effect: Statistically, the death of one spouse shortens the life of the other. I wanted to find out why.

At the time, you were working in poor neighborhoods in Chicago. Didn't you find it depressing to spend so much time with dying people?

No. But eventually I felt myself burning out emotionally. One reason for that was my own aging: More and more, I was getting patients who were younger than me.

Which reminded you of your own mortality.

Of course. The cases also kept getting harder. At first I treated a lot of old people. But then family doctors learned more about palliative medicine. So the only patients who still came to us were those whose doctors had given up on them—the thirty-year-old mother of two children with ovarian cancer whose pain doesn't respond to any treatment.

What can a doctor do to help in a case like that?

Take her needs seriously. We've studied how patients define a good death. Many of the answers weren't that surprising: painless, at home, at peace with God and the world. But their overriding desire is always to be with loved ones. When my own mother died at the age of forty-seven, I was twenty-five. I took care of her. When I asked her whether she was prepared to leave us, she answered: "My son, you can't imagine how hard that is for me." Now I understand the stages of her dying much better. Dying has a lot to do with letting go. People who are dying lose interest in the future, eventually even in eating. But the last thing they let go of is always their connection to their loved ones—their social network.

For all other animals, food is clearly the top priority.

We think of others because, for us, the benefits of living in social networks far outweigh the costs. Just think of how much more productive hunter-gatherers can be when they work together. In developed societies, the more widely interconnected people are, the more whole regions flourish. And what would my own work be without the constant exchange with my dear friend, political scientist James Fowler? We've been working together for ten years and have become close friends.

But can the relationships among hunter-gatherers really be compared with our own?

The differences are only superficial. In our case, about 10 percent of people have one friend, 10 percent have two, 20 percent have three. Only 1 percent have fifty friends. We studied the Hadza, an ethnic group living as hunter-gatherers in Tanzania. Of course, they have a different way of expressing friendship than we do. To determine their ties, we gave them honey and told them that they could share it with anyone they chose. But their social networks and ours have a very similar structure, as we showed in a paper we recently published in *Nature*.

If it depends so little on culture, the obvious conclusion seems to be that the way we forge friendships is innate.

That seems to be the case. Whether you're an outsider or everyone's darling, genes have a great influence on how many connections you make. They also help determine whether you prefer surrounding yourself with lots of friends or having one-on-one interactions. We also discovered that your friends are more genetically similar to you than other people are.

That would mean that the wonderful phrase "kindred spirits" should be taken literally.

Yes, although the similarities are of course not as great as those between actual siblings. Still, it seems that our ancestors' chances of survival in evolution depended on whom they had relationships with.

It's as if there were a hidden magnetism that draws us together to this day. And I thought I chose my friends!

Ten years of research have completely changed how I view myself. Now I see myself more as part of a larger whole—a human superorganism. Its life is far more complex than that of each individual. We've made computer animations of how social networks develop. They're really moving. You see a network that's constantly changing—as if it were a living, breathing thing, even with its own memory. Ideas and germs spread within it. And when a wound opens up in it because a person dies, it heals.

Now you sound almost religious.

This touches on ancient philosophical and even theological questions: What are the origins of love? Why do we have friends? Why does altruism exist? For tens of thousands of years, humans have had only one natural enemy worth mentioning: other humans. So our species evolved in a world in which we could either cooperate or fight. By preaching love, Jesus, for example, recognized that fact. Whatever you think of religion, let alone of Christianity, he was a very smart man.

> But most of us are really invested in our independence, in feeling like the agent of our decisions.

As the philosopher Eric Hoffer put it, "When people are free to do as they please, they usually imitate each other." Think of a buffalo in a stampede. Would that buffalo claim that it was running to the left because it had chosen to? It's doing it because that's what the herd is doing.

> That's what I mean. No one wants to see themselves as a herd animal. Maybe you received death threats for deeper reasons than sensitivity to discrimination: Some people felt as if their sense of self was under attack.

That's possible. But the French sociologist Émile Durkheim already showed, in the century before last, how tenuous our individuality is. Could you imagine a more personal decision than to take your own life? Yet the suicide rate in France has actually depended for centuries on what denomination you belong to: Protestants kill themselves more often. No one is an island. We have to come to terms with that.

> Our attachment to individualism is actually a quite recent development. In the Middle Ages, not many people would have seen what's so desirable about personal independence.

In recent centuries, we've been enormously successful at dissecting the world into smaller and smaller parts. The human body consists of organs, organs of cells, cells of molecules, molecules of atoms. In the same way, we view society as being made up of its smallest units: individuals. And we thought that if we studied the atoms, we could also understand the larger whole. But now it's becoming increasingly clear how much the whole affects the component parts—in biology as well as in our social environment.

> Okay, but why do we need the illusion of individual agency—and the unlimited freedom we associate with it?

If I only knew.

> Maybe so that we feel responsible for what we do. If I believe that I make choices freely, then I also know that I have to answer for all my actions. And then I have a strong incentive to behave morally.

But that way of thinking sets up a false dichotomy. If someone does something wrong, it's not solely because of either choice or environment; it's a result of the interplay between both of these forces, of course. And efforts to promote moral behavior are often more effective when we start with the social network. In Chicago, for example, there's a program that is trying to break the cycle of violence. When someone is shot, social workers immediately rush to the home of the victim's family and try to dissuade the whole group from seeking retaliation. That works much better than the threat of ending up in prison for a revenge murder.

> Most people think of moral decisions as something that each of us makes alone with our conscience.

I believe that morality stems from our social connections. Not only violence, but also fairness and generosity are contagious.

> It's odd that these days we think and feel more individualistically than ever—while at the same time we are networked with more and more people. Have Facebook and the like changed our collective life?

If you had asked your great-grandmother how many friends she had when she was a little girl, her reply probably would have been: two or three close friends, a handful of casual friends, and a lot of loose acquaintances. You'd get the same answer from your daughter. What the network of our relationships looks like can't be changed by technology. But it makes it easier to connect.

> It seems to me that online networks do have an impact on the real world. Just think of the 2011 Egyptian revolution: Would it have been conceivable without the Internet and Twitter?

I don't think so. But the networks conveyed information, not revolutionary passion.

Can those two things be separated? The more I know where and how I can take to the streets with like-minded people, the more likely it is that I will do it.

We will soon know the answer. Colleagues of mine are studying the Twitter data from that time. But despite all our technology, friends we see regularly have the greatest influence on us. We've investigated that on college students' Facebook sites. We thought that their taste in music and movies would spread through the network in the same way we've seen elsewhere. But that wasn't the case at all: As a rule, one Facebook friend's fondness for the Killers or *Pulp Fiction* increases the likelihood of another's only when there's a photo of the two of them together online—meaning they know each other personally.

But we typically enjoy movies and music with friends. Other preferences might be more likely to spread via the Internet. And some communities exist only in the virtual world, like fan groups of online games.

Of course how something spreads always depends on the contagion itself. Some of my colleagues are currently trying to seed Facebook and Twitter with various ideas to find out which of them spread best. The problem is that we get only the publicly accessible data from Facebook.

I find it hard to accept that extensive knowledge about the lives of billions is in the hands of a few companies with unchecked control over it.

That will change. When energy and water companies covered our countries with their networks a hundred or so years ago, they too could do whatever they wanted. Now they're heavily regulated. The same thing will happen with the Internet companies. Laws will dictate what data they're allowed to collect and store, how they have to disclose information to their users, and what they have to divulge.

Will that suffice? It's possible that we're still in such an early stage of electronic networking that we don't even suspect the consequences.

You might be right. It worries me somewhat that ever more precisely targeted advertising messages from the Internet have been doing even more to fuel human greed. But unsuspected opportunities present themselves as well. Imagine you had asked a social scientist twenty years ago what the ideal research instrument would be. He would have replied, "A fleet of tiny invisible helicopters." A helicopter would hover over every person, recording every word, every movement, every desire. Today, with the Internet, we have that.

The Software of Life

Biochemist
Craig Venter
on the human genome

◆

CRAIG VENTER MIGHT BE THE MOST controversial scientist of our time. His admirers praise the charisma, courage, and intelligence of this man who has decoded more genes than anyone else. For his enemies, however, he is "Darth Venter": Just as the brutal tyrant Darth Vader in *Star Wars* defected from the good guys and went over to the dark side, so too did Venter leave the National Institutes of Health (NIH), the United States government's medical research agency, and establish private research institutes and companies in order to become the sole ruler over the universe of genes.

Venter was born in 1946 in Utah and spent his youth in California. After military service in a field hospital in Vietnam, he studied biochemistry, became a professor at the State University of New York at Buffalo, and later joined the NIH.

We first met in 1999. At the time, Venter had just declared that he would sequence the human genome in only three years with a newly founded company, Celera. That announcement set in motion the race for the human genome and gave rise to unprecedented public euphoria about genes: A widespread perception arose that genes are the key to the fate of each human being.

This time we met at Venter's summer house on Cape Cod. Venter now presided over a private research institute known as the J. Craig Venter Institute. I wanted to know what had become of the genome project, the hopes and fears it had unleashed—and Venter himself. He appeared wearing jean shorts and black sunglasses. He took off the sunglasses only toward the end of the conversation.

◆

You're one of the first people to come to know your own whole genome.

Well, let's correct that. I was not one of the first; I was the first.

What about your old rival, James Watson, who discovered the structure of DNA?

What about him? He was second. Watson's genome was published this year and mine was published last year, even if you're talking about the complete genome. But Watson's is not a complete diploid genome. My genome is the only one that is completely sequenced.

Did your genome teach you anything new about yourself?

Genomes can't tell people very much about themselves. They can tell people about some specific traits. Drug metabolism is perhaps one of the most interesting, because things like caffeine metabolism are very much dependent on genotype. I have two copies of very fast metabolizing genes, and so I can drink lots of coffee and it has no impact. People who have slow metabolism or even medium metabolism for caffeine have a greatly increased risk of heart arrhythmias or heart attack if they have multiple cups of coffee, so it helps explain all the confusing information in the scientific literature about caffeine's being good or bad for you. It totally depends on your metabolism.

Same thing with alcohol, probably.

That's been much more known just because more people do that experiment, and people without the ability to induce ADH [the hormone that increases water reabsorption in the kidneys] can't tolerate alcohol. Also, those genes associated with the alcohol flush are much more common in Asian populations but occur throughout different populations. I don't have those genes, and I have good inducible alcohol dehydrogenase. So I drink lots of alcohol and lots of coffee. [Laughs] Usually in that order. [Laughs]

As you describe in your book *A Life Decoded*, your genome told you a bit more. For example, you have quite a good chance of reaching ninety years old, but you do have an increased risk of developing Alzheimer's disease. Now, how do you deal with that knowledge?

The same way you deal with the rest of the knowledge. It's an interesting fact that, again, nobody really knows how to fully interpret in the sense of a genome, because you can't look at any one factor and say, "That's the risk." People who smoke have ten times the risk of Alzheimer's disease, so things are not just genetic, they're environmental. We now know from lots of studies that the brain is like any muscle in the sense that if you exercise it and use it constantly, your chance of having dementia is much lower. So I think being an active scientist, actively taking statins, trying to have a healthier lifestyle—all these things, plus family history—I'm not overly concerned about it, but obviously I looked at all these different issues. But that's why we have to understand the complete human genome. We can't have just partial ones.

Don't you mind being, in a sense, genetically naked in front of the whole world?

It hasn't made any difference. In fact, one of the things I applaud Watson for—given all the negative approaches in the US government that are totally against public disclosure of genetic information—Watson, like myself, clearly seems to believe that it doesn't matter, that human life is not all genetically determined, and that the more scientists and leaders in the community who make their genotype and phenotype information available, the more it will help the rest of the world interpret the information. So I cannot anticipate any adverse consequences from doing so. In fact, we need to encourage more people to do the same thing.

> Do you happen to be familiar with the autobiography of Giacomo Casanova? He is known for the women of course, but he was also extremely risk- and sensation-seeking—he escaped prison in Venice, and he gambled. Now, his risk-seeking behavior has been linked to his dopamine receptor genes. Do you find this possible?

No. The assumption that the extent of human behavior will be caused by a minor change in the dopamine receptor—I think it's extremely naive. That complex human behavior is caused by point mutations in genes that are common to multiple neuronal pathways, I think is extremely unlikely. We call them light traits, or even larks, because they're curious or funny, but I think things will not correlate in that way in human physiology just because somebody published it as a paper in the scientific literature. The same thing happened with the so-called "gay gene" that never materialized—it was never reproducible. It doesn't mean that there may or may not be genetic changes associated with sexual preferences, but it's not going to be trivial like a single point mutation. Casanova's behavior is certainly not going to be explained by one base pair change in his genome.

> But it has been customary to ascribe literally all variants of human behavior to genetic causes. My favorite is a paper on the effects of social class and race on pair-bonding, which claims that an alleged upper-middle-class preference for cunnilingus and fellatio can be traced to the workings of evolution.

The scientific community hasn't exactly been right too many times about the human genome. The estimations were 100,000 to 350,000 genes. It turned out there are only 23,000 genes. We have a relatively naive scientific community when it comes to anything about human genetics, human variations, the cause of traits, the cause of diseases. And that's the point of doing things in the scientific fashion: getting multiple complete human genomes, on the order of tens of thousands, along with complete phenotypic information. And once we have that, and all the stuff has been published, it'll just be viewed as crap that nobody pays attention to again. Ninety-nine percent of what's published in the scientific literature never gets looked at twice.

Although still we read about things like the decoded human. You have made headlines for this of course.

I think there was a lot of naive optimism at the time, but I think historically having the first decoding of the genome, and with Watson and others following along, this will be a very sharp turning point in the history of humanity. It will be remembered, even though the ways people think about it now and thought about it in 2000 were extremely naive compared to how they'll think about it in three hundred years. But they'll still remember this time, because this was the first time in history when we're able to go from no understanding, or very limited understanding, to at least having all the information. So, yes, it will change the view of humanity, but not instantaneously. Trouble is, we're still trying to accurately interpret the information. You might recall that I said that the genome race was a race to the starting line, not a race to the finish line.

I'd like to get into the matter of the complexities of the genome, on which you gave a lovely and astonishingly simple example in your book: the color of your eyes. Let's elaborate on how such a simple feature, once thought to be determined by one or a very few genes, can be much more complicated than people thought it would be.

Well, it's a notion that we must be simple machines. If you're making a robot that looks like a human, you select the color and say, "That's the color of the eyes." But what seems like a simple binary thing when you're building a machine is complicated when you are dealing with biological determination. Things like eye color or hair color or face shape aren't simple binary notions. They're complex components that go together. You have one hundred trillion cells that are starting with one cell—that's a lot of cell divisions, and there is a bunch of events that happen in cell division that, even with the same exact genetic code, you'll never get the same answer twice. You have genetic variation. You have built-in components that are affected by what we eat, what we consume, and what we're exposed to. So it's not surprising that there are multiple genes, multiple cell types, and multiple components that go into what we see as a simple term that we call eye color. We're very much determined by and limited by our visual acuity.

What do you mean by that?

When people look at each other they don't see one hundred trillion cells. You don't see 23,000 genes and constantly varying dynamics inside each one of those one hundred trillion cells. We see simple things, like heads and eyes and noses and mouths, and think that these are simple features because that's what we see and that's what we measure each other by and even judge attractiveness and intelligence by, because we have nothing else to go on. So science is the ability to get beyond the limits of our own capabilities, by using tools.

What do you think of the metaphor of DNA being the Book of Life?

I never liked that analogy. I never use the Book of Life in our report. Software I like maybe, but . . .

What is the difference between the Book of Life and software?

Well, if you read a book online, you're ignoring the software that's behind it, that makes the book possible. You're just reading the end result. A book is unchanging. It's a snapshot of a point of view. So the software is what runs things, not what people look at. When you look at your computer, you don't see software. You're not looking at the code behind things, you're looking at the pictures or you're looking at the text. If you read a book online, there are maybe a million lines of code that your computer is using to present those images to you. So that's why I see the analogy with DNA as much more the software, the book analogy as much more the phenotype.

Now, the question that has always puzzled me is: How can it be that a bunch of information software makes people predisposed to learn language, for example, at a certain moment in their life? How does it shape us? These are such extremely complicated things, and maybe the really interesting questions and answers are in developmental biology. At least you have to tell the story of a life to understand what happens.

Yeah, but I think those complexities will be understandable in the future. In adults there is a pretty standard test you can give to find out if they have language abilities or not. So it gets down to fundamental neuronal wiring, which is a combination of genetics and stochastic events. There's clearly a genetic component to every human trait that we measure and cherish. So intelligence, personality—there's a lot of hardwiring that goes into these things.

That brings me back to risk-seeking behavior. You've sailed into a storm on your way to Bermuda without having listened to the weather forecast.

I personally didn't consider that so risky. See, I don't jump out of airplanes, I don't hang glide off the top of buildings, and I don't climb up the side of buildings without any rope. Those are risk-taking behaviors.

Okay, then I'll just shift my choice of words. Where does your self-confidence come from?

Well, again it comes from learned experience. I think I was relatively low in self-confidence because, number one, I hated school, and as a result I did very poorly in school. The thing that first taught me self-confidence was taking part in competitive sports. I was a competitive swimmer, and I discovered that by learning some things and applying myself, I could do very well. I remember very distinctly that having a huge impact and changing my self-confidence in my ability to do things.

So I can shift my question again. Where does your competitiveness come from?

I'm sure it's a combination, like most things. It could be simple birth order. I'm a second child. Statistics show second children are quite often much more competitive than the firstborn. I don't know. This is not my field of expertise, but it's sort of like saying, "Where do type A personalities come from?" Right?

I don't think they exist.

I know they exist.

How do you know?

I live with one. I've seen many of them. Do you have any children?

Yeah, I do.

Okay. I assume you noticed from day one they had unique personalities, if you have more than one.

Oh yeah.

I mean there's a certain . . . there's a substantial amount of hardwiring in humans, something I don't think our society is totally ready to face. But there's a big distinction between having hardwiring for many traits and genetic determinism. Somebody may have an aggressive personality, be very competitive. They can end up an Olympic champion. They could end up a criminal, right? It doesn't determine life outcomes. It determines the set of operating tools that each of us has or doesn't have. Then a lot of parents say that their biggest question is, "Can we actually influence the outcome of our children, or are they basically hardwired to be who they're going to be?" Obviously they can influence their children by providing love and nurture in relationships, being positive role models, helping them to experience a lot of things. Because while we are one of the most genetically advanced species, and we're genetically hardwired, we're also probably the most plastic species.

Now people are learning to understand how the workings of genes can be changed by the environment, of course by learning, but also by stress, and so on and so forth. And for better or worse, these changes can be passed on to the next generation.

Well, it becomes very difficult to accurately measure what gets passed on to the next generation through genetics versus environment. I look at my fiancée and see how similar she is to her parents in many ways. Is that genetics, or is that because they spent twenty-some-odd years together? It is very difficult to separate the two in terms of complex behaviors. I mean, these are all great questions. They're not necessarily things we can answer today, but we know we need to try and reconsider some of these issues.

Do you think you would have become highly successful, and of course a highly competitive scientist, if you'd grown up, say, in the Rockefeller family? If you were adopted, or somehow switched at the hospital, and taken into their family?

I think there are so many different influences—it's obviously impossible to answer that question. I think one thing that has made me and makes other people highly successful scientists is just a basic innate curiosity. You know, asking questions about life. I attribute that to maybe doing poorly in school, and so I didn't have creativity drilled out of me like people who have either forced themselves to be good at, or are naturally good at, memorization. I would hope that if I were in a different environment I would have the same innate curiosity. So I think there is some general categorization that seems in a broad sense to be holding up in terms of personality effects, that seems to be genetic. People are either optimists or pessimists. I'm obviously an optimist. If you're an optimist and curious, I think anybody that's in this category would have the makings of being a great scientist.

This brings me back once again to self-confidence. I read about your time in Vietnam, and there's one among those stories that I found hard to believe—the one with the sea snake. I mean, if it touched you on your leg, how could you have remained so collected? How could you have known that this thing was a sea snake at all?

Well, when you're swimming and there's surf, you see sea snakes swimming into the surf all the time. The odds are pretty good that if something hits you in that environment, it's a sea snake.

Isn't that risk-taking, swimming into sea snakes?

Well, risk-taking in a war zone is very different from doing the same thing at age sixty-one.

But the sea snakes have nothing to do with the war, right?

Well no, they have everything to do with the war, because in a war zone you're dealing with death every single day. Life became a very cheap commodity in war. That's the whole basis of wars. The governments are willing to expend lives of their youth to accomplish a political end. Not too many people can deal rationally with ten to a hundred people being killed every day, and you have to deal with that. Obviously, you have to compartmentalize in some way, and everyday life is a risk, and so swimming with sharks and sea snakes and sailing with people shooting at you didn't really seem like risk-taking behavior, because you could just sit there and be rocketed as well.

Reading your autobiography, I was kind of surprised that you became a scientist at all, having worked for so long with that adrenaline receptor—the endless monotonous experiments. . . .

Science is about ideas and it is about asking questions and devising ways to get those answers. And so there was nothing at all—even when we were trying to isolate the adrenaline receptor—that was boring. It was frustrating. Lots of experiments didn't work, but there is truly nothing more exciting than having an idea, thinking about a way to test that idea, and getting an answer that gives you new information about the world around us. It's the most exciting thing that I do, and there's nothing that I'd ever trade for it—starting off with my thesis about how adrenaline works in the heart, being able to work it out, and solve these things when it's something nobody did before I did those experiments. Even if 99.9 percent of the world doesn't care about it, those were very exciting things to do. Behind every science experiment, and behind sequencing the human genome, there's a lot of tedium.

Extreme levels of tedium?

Well, not extreme, because what we did is set up procedures to make it not so tedious. We automated everything. In fact, the biggest complaint that I've gotten from my colleagues who worked on the [sequencing] team is that it's a shame we did it so quickly and we didn't have time to enjoy it more, because it was such a phenomenal time period. Most science crawls along.

Do you think that's unfair, the fact that, in public perception, you are Mr. Genome, despite having so many people working with you?

Of course it's unfair. Tomorrow nobody is going to remember who was second or third at the Olympics either. Nobody is going to remember the five or ten thousand people behind the Nobel Prizes that come out of high energy physics, right?

What does fame mean to you?

Fame is one of those things that other people perceive that you have. There's very little actual utility other than enabling avenues that may not have been possible before. But my personal perception is that I don't feel like a famous person. It's one of those attributes other people apply to you. So what I find is, it's a wasted phenomenon unless it's used to create more good. I think my prior successes are very helpful with what I'm trying to do now.

Thinking about what lies ahead now, one of the next steps to better understand genotypes and phenotypes is to sequence many people.

I've said we need ten thousand. When we know what genetic component it is that is associated with things, we'll know absolutely what's genetic and what's environmental. And, I think by definition, causal.

Yeah, but you would not know what the genomes do. Not yet, right?

That doesn't matter. What we want to do is have predictive science. You might find genetic changes like people have associated with diseases like Alzheimer's disease, and you might find a totally different pattern in people that have those genetic changes but never get Alzheimer's disease. If these patterns give you a truly statistical, predictable outcome, then that's what you need to know. And if the correlations aren't high, it means there are other factors, including environmental factors.

> If you believe there is a genetic predisposition for, say, your liking the water, I would find it extremely interesting to learn how it operates. But in order to learn that, you would first have to learn how the brain operates.

First, you have to prove whether there is even a genetic correlation. There may not be.

> Of course.

We should be able to answer that, but first we need to know how to digitize the information of whether people like the water. I knew of a person, when I lived in Buffalo, who said they'd never seen the ocean, and they didn't really want to because they didn't think they would like it anyway. I mean, there are so many different explanations for that. Probably the difficulty would not be the genetic correlations, it would be the measurement in the first place. How do you measure and score whether somebody really finds peace from looking out at water or doesn't, you know? Is there a genetic component to liking the mountains, liking the sea, liking being in the water? Maybe there's not. Maybe it's all learned behavior.

> Sure.

If you find the correlation, then you can go back and maybe try to understand eventually the normal pathways and causality, but that's the point of reductionist science. Almost everything right now in human biology is guesswork and supposition, and ritual-istic behavior.

> I totally agree, but I think the jury is still out on whether reductionist science in a human or in any complex organism works at all.

Well, obviously it has worked numerous times. But the existing limits of genetics have been exhausted, and so we need new tools to go further. Reductionist biology has been applied to humans and human diseases. That's why we know how to treat hypertension; that's why we know how to diagnose different ailments; that's why we know the genes associated with Huntington's disease.

> But talking now of a decoded human is ridiculous; would you agree with me on that?

No, it's not. What we said is the human genome has been decoded. We have it digitized in the computer. Do we understand it all? No. So it's not ridiculous to say it at all; it's a fact. We've decoded the human genome. We have not decoded humanity or humans or Books of Life or any of these other metaphors that people use that I think are silly to ridiculous. But we have read the human genetic code.

> You haven't stopped there. You've turned to synthetic biology, assembling artificial organisms. I've asked myself why. There are so many things about the workings of the genome that are not understood. So what's the point of assembling genomes at this time? Isn't it much too early for it?

We've shown it works. We published the first synthetic chromosome a few months ago, so it's clearly not too early. But it took over ten years of really hard work to come up with new methods to enable it.

> Yes. But now you have the idea of creating microorganisms that do certain things.

Yes.

> And in order to do this, don't you have to understand the workings of this "software" much better?

If we just take any one milliliter out of the river here, or out of the ocean, there are a million bacteria per milliliter and ten million viruses. Most of the genes, most of the DNA and their functions, are yet to be discovered. In fact, most of the species have never even been seen before. So biology is very much in the empirical phase. People like to pretend that it's not, and that it's all hypothesis-driven. It's in the same vein of discovery as in the sixteen and seventeen hundreds, of first looking through microscopes, of people sailing around and discovering the lands, of Darwin going through the Galapagos coast and seeing the species there and observing them and describing them and building those into a very important theory. We're describing the world around us. We have to base things on empiricism, and we have to create new tools that allow us to expand the capabilities of individual scientists by millions of times. So that's what we're doing.

The Poetry of Molecules

Chemist and poet
Roald Hoffmann
on beauty

◆

"DON'T MAKE THE CHEMIST LOOK TOO STIFF," admonished the photo editor. "Don't worry," replied the young photographer assigned to accompany me on my conversation with Roald Hoffmann. "He's a poet."

She was right. In Hoffmann's office at Cornell University in rural upstate New York, there's not much to remind you that a world-famous scientist works here. Indian masks and a statue of the Hindu god Krishna playing the flute adorn the room. There are pinecones and editions of the Talmud lying about. From the ceiling hangs a net made of feathers. "An Indian artist from the area made it," Hoffmann explains. "It's a dream catcher."

Hoffmann was born in 1937 into a Jewish family in a town near the then-Polish, now-Ukrainian city of Lviv. He survived the German occupation hidden in an attic. After the war, he studied chemistry at Harvard. He was not even twenty-seven years old when he made his first groundbreaking discovery. With his colleague Robert Burns (R. B.) Woodward, he found rules with which chemical reactions could be predicted. That earned him a Nobel Prize.

Scientists are fond of pointing out the quantity of their publications. Hoffmann's list is five hundred titles long and keeps growing. It includes not only scientific articles, but also essays on beauty, art, Jewish intellectual history—and four critically acclaimed collections of poetry. At the time that I meet with him, Hoffmann is at work on his third play.

◆

Professor Hoffmann, do you have a favorite molecule?

Hemoglobin—the red pigment in blood. It's a molecule of truly baroque magnificence. About ten thousand atoms, mostly hydrogen and carbon, are bound into four chains that coil around one another. The whole thing looks like four tapeworms making love.

Pretty convoluted. . . .

Yes, but only at first glance. In reality it's a mixture of disorder and order—for most of the curves actually have a purpose. Four disks, known as hemes, are wedged between the twists of the chains. Right in their center is a lone iron atom. That's where the red color comes from. The oxygen we breathe binds to the iron. Each heme takes in one or two atoms of oxygen.

Ten thousand atoms to package all of eight oxygen atoms? What a waste.

But beautiful. Don't you think?

Women can be beautiful; ice crystals are beautiful. We can see them. Hemoglobin, however, isn't even visible under a microscope.

You can't see music either, but it's still beautiful.

We can hear it. Since antiquity, philosophers have considered sense perception to be central to our experience of beauty.

A far more decisive factor than sense perception is what interest an object arouses in you. The experience of beauty comes from a tension between your mind and the object. But you're right: The interest has to come from somewhere. It begins with sensual attraction.

To an attractive body maybe. But chemistry? I imbibed your science with my mother's milk, so to speak. My father was a chemist, my mother and grandmother were chemists, and even my great-grandfather was the director of a chemical research institute near Vienna. And yet, among all the sciences, chemistry is the one that least captivates me.

Did you have a chemistry set as a little boy?

No.

You see? You were missing the sensual side. Chemistry is interesting because it smokes, bangs, and stinks. That's where the attraction comes from.

That youthful attraction turns later into a very intellectual pleasure. By contrast, when a painting or a sculpture appeals to us, we usually don't think much about it. The experience is immediate—our heart wells up with emotion. Later an intellectual relationship with the work might develop, but it's not necessary. So can a molecule really be beautiful in the same way as a work of art?

The criteria are different. In art, emotion plays a greater role, and in science, intellect does. Take a look at the picture over my desk. It shows an idol from the Cyclades, five thousand years old. When I look at that marble female body, I don't think much about what influence Egyptian and Cycladic art had on each other. Just seeing the statue gives me a warm feeling. But now look at the ecstatic woman's face in the picture next to it. It's Saint Teresa of Ávila, a sculpture by the baroque artist Bernini. There the intellect plays more of a role.

Teresa appeals to me not only because of her appearance, but also because she, a Christian nun, had a Jewish grandfather, and because women's visions interest me. Ultimately, that was the only way women could express themselves in a male-dominated Church. The sculpture tells me a story. There's a tension between the work of art and me.

Hemoglobin . . .

. . . tells a story too. The chains are intertwined in such a way that a sort of pocket forms between them, into which oxygen in the lung can slip perfectly. When the passenger has taken its seat, the molecule changes its form, actually encouraging a fellow passenger, another oxygen molecule, to enter a neighboring pocket. And a third. And a fourth. Then the door snaps shut. As a result, the color changes: The blood turns bright red. In the brain or in a muscle, the hemoglobin releases the oxygen by assuming its previous form. That's why the blood in our veins is a purplish red. The way this molecule travels through the blood vessels while constantly transforming is as thrilling to me as the story of Odysseus.

But that kind of beauty is accessible only to the few. Anyone who views a painting or listens to a piece of music can enjoy it. To find pleasure in the beauty of hemoglobin, you have to have studied chemistry.

The two possibilities don't have to be opposed. After all, I never claimed that the beauty of molecules is greater or more important than that of a work of art. But beauty can also exist in realms where we don't usually expect to find it—in scientific research, for example. And in recognizing the fact that the deep understanding of a molecule can also stir an aesthetic emotion, we see science in a new light. It seems more human.

Some scientists are guided in their research by the search for beauty. Albert Einstein, for example, felt really uneasy when he found an equation ugly. Truth in nature, he thought, was simple and beautiful.

I don't believe in that. The world is complicated. Why should nature have a tendency toward simplicity? It's only our mind that seeks simplicity, because it can cope with it more easily.

You seem to find pleasure in complexity. But simplicity, too, has its magic. Don't the perfect proportions of the Parthenon appeal to you?

Or molecules that look like perfect cubes? I used to idealize things like that. But the older I get, the more fascinated I am by complexity. It might also have to do with our time. There are epochs like Greek antiquity that favor simple forms. In other historical periods, however, the complex is regarded as beautiful. That was the case in the baroque era, for example, and it's the case today. Many people find fractured architecture, such as Frank Gehry's buildings, far more beautiful and interesting than the Bauhaus cubes of the postwar years. I, for one, have had my fill of simplicity. It doesn't tell a story.

Why do people experience beauty at all?

Categories like beautiful and ugly are in part genetically determined. It could be that people originally found beauty in what was useful. So our ancestors might have felt attracted not only to particular edible plants, but also to all living nature—for no species can survive on its own. I imagine this is the reason that the pleasure in the living, in irregularity, defines our sense of beauty to this day. That's part of why we prefer flowers and wood to plastic.

Genetic programming of that sort might exist. But it would hardly explain what fashion or what music we like. There's nothing natural about the notes of a string quartet or an electric guitar.

With the development of language and culture, the sense of beauty of course became much more complicated, and it can no longer be explained solely in biological terms. People today have learned many aesthetic judgments in the course of their lives.

And yet we can agree surprisingly often about what's beautiful and ugly. Everyone admires the *Mona Lisa*.

But that's precisely because we can no longer look at it impartially! Everyone has seen the painting thousands of times and heard or read countless judgments of it.

But the mystery is how something like the *Mona Lisa* became so famous to begin with. The first people to see Leonardo's work, five hundred years ago, already praised it. Besides our affinity to nature, there must be additional principles by which we judge beauty.

Let's come back to the question of complexity and simplicity. Our mind is programmed to look for patterns. It favors simplicity. We feel at ease when we immediately understand something—whether it's a painting, a building, or a molecule. But then the thing quickly becomes boring. We need something more to keep our interest.

Are you familiar with the Park Güell in Barcelona? There's a huge terrace there supported by columns over a hillside. It's designed by the architect Gaudí. At the edge of the terrace is a bench that curves away from the hill and then toward it again in a completely regular wavy line. That is simple. You understand immediately the form of the terrace. . . .

And only in that way can there be sensual attraction. If the first impression were too complex, we would be scared off.

Possibly. But that's just the beginning of the story—for the bench is covered with multicolored ceramic tiles in a completely irregular mosaic. There's no discernible pattern. Here you have complexity. The sizes and colors of the tiles seem to be assembled randomly, yet in a pleasing way—the artist/architect's métier at work. We don't find order or disorder by itself aesthetic. Beauty comes from tension: between order and disorder, simplicity and complexity.

We experience beauty where there's still a mystery to solve. And we have to believe that we can solve it.

Kant was deeply mistaken in that respect. He asserted that beauty was "disinterested pleasure." In his view, we can judge as beautiful only what isn't bound to us by any intention.

According to Kant, I couldn't find a woman beautiful and desire her at the same time. By the way, I've never thought Claudia Schiffer was especially beautiful or attractive. I'm more partial to someone like Juliette Binoche. . . .

Because you sense mystery in her.

Our experience of beauty is based, in your view, on interest and use-fulness. Therefore, it's a form of desire—a longing to solve the mystery. Perhaps the greatest works of art are those that arouse that yearning but never fulfill it.

Yes, but there's more to it than that. For me, the pleasure of visiting a museum lies in feeling my physical senses and my mind interact in response to a work of art. I experience the unity of my own inner world. Even more than that, I feel connected to everything that surrounds me. And I'm reminded of the good side of human nature.

Can something horrible be beautiful?

Think of Goya's etchings on the horrors of war. He shows mutilation, killing, torture with unprecedented precision. The works are masterful—certainly a borderland of beauty. I find them beautiful.

You yourself have written poetry about your experiences under the German occupation.

I've often been asked whether I want to publish those poems, currently scattered over four collections, in a single Holocaust volume. I've always refused. Those experiences belong together with all my other ones—with my poems about love or chemistry.

One of those poems is titled "June 1944" and deals with the time after your liberation by the Russian troops. You describe yourself in the poem as a six-year-old who, in his hiding place, has forgotten what wind is. There the boy looked out through a hole in the wall at playing children, whose "giggles / bounced in, but no wind, / for the brick hole was small."

A Ukrainian village schoolteacher hid us: my mother, two uncles, an aunt, and me. I was the only child. Crying in that hiding place would have given us away. I learned not to cry. My aunt's child was only two at the time. The child, whose crying would have betrayed us, was given away to a Polish family and ultimately murdered by the Germans. My uncle had a gun in our hiding place. If the

Germans had found us, he would have shot us and himself. But I can't remember whether I knew that at the time or my mother told me later.

Where was your father?

In a labor camp. But in those camps there weren't many German overseers. The guards were mainly non-German collaborators. They could be bribed with cigarettes, chocolate, or whatever. And since my father was a civil engineer, he could move rather freely in and out of the camp; he was of value to the Germans—he had built some of the bridges the Russians had blown up in retreating.

Why didn't he use his freedom to join your family?

He could have. But he used his freedom to smuggle weapons into the camp. They planned to break out in a large group and escape into the forest until the Russians arrived. If all had gone as planned, he would have joined us. The breakout failed, and the guards killed him. He was a hero.

You survived the Holocaust against all odds. "Eighty of 12,000 Jews in our town survived," you have written. How do you feel when you hear German today?

I have no trouble with Germany. When I'm there, I sometimes wonder what some of the older people were doing during the Nazi era and never told their children. On the other hand, my scientific work has found a special resonance in your country, and as a result, many young Germans have come to my group as research fellows. Over time, some of them have become like family to me. So I've developed new, strong ties to Germany. By the way, in those days we—especially my mother—felt a much greater antipathy toward the Ukrainians. Ultimately, they were the ones we feared would betray us. Even though the murderers were Germans, of course. Crazy, right?

Is the memory of the danger still vivid for you?

Certainly. And it leads to strange reactions: In restaurants I'm afraid of waiters, because they wear a uniform. And to this day, I can't stand in front of a window at night, because when we were in hiding, the threat came from outside. Of course, after more than sixty years, my memory is buried under many layers. That's why I traveled to the Ukraine last summer. That was the first time I had gone back to the town where I was born and visited our hiding place.

What was that like?

The attic was bigger than I remembered it. Because it was really cold up there, we spent the second winter in a room on the ground floor. When Germans were in the vicinity, we crawled into a hole, a bunker we had dug under the floorboards. While we were huddled there, we sometimes heard the soldiers' boots over our heads. That room is now used as a classroom. And do you know what's hanging on the wall? The periodic table. It's a chemistry classroom. And under the periodic table is a quote from the Russian chemist and poet Lomonosov in Ukrainian: "Chemistry spreads its arms wide for the good of mankind."

That sounds so improbable that it's hard to believe.

For me, it was a shock when I was ushered into the room. Of course, the schoolchildren had no idea what had once happened between those walls.

Do you believe in fate?

No. But sometimes it's hard not to. Scientists are no different from other people in that regard. They know well that after the roulette ball lands on red five times in a row, the probability is no greater than usual that it will land on black the sixth time. The ball has no memory, after all. And yet in a casino, they'll still bet on black—when no one's looking.

Why did you become a scientist?

It was an accident . . . or at least I didn't feel any special calling to chemistry. Following my mother and stepfather's wishes, I was unenthusiastically preparing to go to medical school. During semester breaks, I had jobs in research laboratories. I liked that work; I had done chemistry experiments as a young boy. So I went into chemical research—though I was actually flirting at the time with art history. I had attended a few courses in art and literature, and a whole world had opened up to me. But I didn't have the courage to tell my parents. Those were hard times for immigrants; my stepfather was unemployed. So I stuck with chemistry.

Do you regret it?

Sometimes. On the other hand, I enjoy chemistry, and I think I have a lot to offer my field and especially my students. And I do have the opportunity to express myself artistically, even though I didn't begin writing poetry until I was forty.

How do you make the switch from scientist to poet and back again?

I have to go to a different place, ideally a natural setting. It takes me two days to put science behind me; during that time, I'm often plagued by headaches. After that I can write roughly a poem a day.

Writing and scientific research have a lot in common: You choose a subject and try to go where no one has gone before.

Whether words fit together into a whole or a connection becomes apparent in nature, I have a similar experience of that wonderful moment when suddenly everything clicks. But the paths to get there are somewhat different. In poetry I usually proceed from the tension between a few words and begin to play with them. I have no idea at the outset what's going to come of it. What course a research project will take is usually clearer from the beginning. It's sort of like a game of hide-and-seek with nature. It resists giving up its secrets, and yet at some moments it reveals them. And in the end there's a sense of liberation when you finally have what you longed for—until the next challenge arises.

What did your scientific colleagues say when they found out that you spent a portion of your time writing poetry instead of doing research?

In the beginning, some of them teased me: "You have the luxury of spending your time writing, we don't." They didn't know how much harder it was to write poems. Or to publish them. In the world's best academic journal for chemistry, about two thirds of the longer articles and a third of the shorter ones submitted by researchers are printed, whereas in even an average literary magazine, under 5 percent of the poems are published.

And how do writers and artists react to you, the intruder from the field of chemistry?

Occasionally I argue with authors who claim that science is like dissecting an eagle: Afterward you know all about the bird's internal organs, but it can no longer fly. That attitude is based on a perception of science from the nineteenth century, when scientists really did dissect every living thing. But that doesn't have much to do with modern molecular biology, for example; incidentally, those same critics have no trouble eating chicken or turkey. Of course, it's not necessary to understand a bird's metabolism or the aerodynamics of its wings to experience the poetry of its flight. But it doesn't hurt either. On the contrary, more knowledge about nature opens up new ways to experience the magic of reality. Think of the beauty of hemoglobin!

Does it bother you that scientists are often regarded as emotionless rationalists—Mr. Spock in the laboratory?

It certainly does. But scientists themselves are to blame. First of all, they describe their research in an atrocious style, in which everything personal is left out—as if the work were done by machines. That's a German inheritance, by the way: To distance themselves from the nature description of Goethe and the Romantics, German scientists, in the first half of the nineteenth century, developed a way of writing that excluded the researchers themselves and anything poetic. And the rest of the world took on that wooden idiom and uses it to this day. As if that weren't discouraging enough for outsiders, scientists also imply that they're supersmart. Which they're not.

The Nobel Prize winner is telling us he's no smarter than the rest of us?

That's right.

So, then, what distinguishes scientists in your eyes?

First and foremost, curiosity. But other people experience that too. Scientists, however, are part of a collective undertaking. They're members of a social system that puts curiosity to use.

Science is like an extremely complex puzzle that hundreds of thousands of people are working on: It's enough when each person contributes a few small pieces to the bigger picture.

Exactly. And for that, no one has to be brilliant. If a scientist wants to solve a particular problem, he can draw on what others before him have published. He can inquire with colleagues. And finally— this is very important—he receives praise when he himself publishes his solution, even if it was only a very tiny step; that spurs him on. Science consists of an endless number of such tiny steps.

And yet the result is often something that changes our lives. The rules that you and your colleague R. B. Woodward established opened up completely new possibilities in organic chemistry. Suddenly it became possible to produce substances no one had previously imagined. How many of those substances would the world be better off never having seen?

You're asking about my responsibility for explosives, failed medications, and poisons? It's not that simple. Woodward and I changed the way chemists think. We showed them connections no one had seen before—such as how the insights of quantum physics could be used to predict a reaction. We gave them a graphic language with which scientists could express those connections very simply. And I've done many other things. Not practical ones, I admit; I don't have a single patent to my name. My work is that of a teacher, not an inventor.

But others have used those insights for inventions.

I'll give you an example. Recently, a colleague gave a talk here about a new drug to fight nicotine addiction. There's a huge market for that, about a billion dollars per year. The drug is a rather simple molecule of twenty-five atoms, but very intelligently assembled.

The production process consists of about ten steps. For two of those steps, the Woodward-Hoffmann rules apply. But our rules couldn't have told our colleague how to take those steps. They told him only what definitely would not work. About twenty of fifty conceivable reactions could be ruled out from the start. So we saved that chemist a lot of work. And yet we didn't take that path, he did. If a drug really comes out of it, I won't earn a cent. How great is my share in the invention? It's impossible to say. I think a tenth of a percent would be enough for me.

Do you really believe that you bear no responsibility for what happens with your fundamental research?

Anyone who brings something into the world bears responsibility for his creations. Even a poem can hurt people, say, when a former lover reads it. Unfortunately, in science it's simply impossible to pinpoint the exact contribution of an individual to a final result. Only rarely are the circumstances as clear as in the invention of chlorofluorocarbons. The chemist who first produced those substances was certain that he was benefiting the world. Those gases are nontoxic, they don't burn, there are no poisonous by-products of their production—an ideal refrigerant and propellant, or so it seemed. Only decades later did it come to light that those supposedly nonhazardous gases are destroying the earth's ozone layer.

You have been unusually successful as a chemist. What differentiates you from your colleagues?

Maybe a better ability to empathize with other people. I've always had a really good sense of what difficulties my colleagues in the lab are facing—even if they haven't verbalized them. And I've then solved those particular problems. This special gift of empathy might come from my wartime experiences. A strong desire to please is often found among people who went through horrible things early in life. A child whose father is killed, or even children of divorce, blame themselves for the evils of the world. They want to show that they're good kids.

Others who had experiences like yours despaired. What sustains your optimism?

Every smile on the faces of my grandchildren strengthens my hope that they will deal with climate change, even if I don't know how. That's the same thing I find in art and science: Both encourage my belief in the inexhaustible inventiveness of the human spirit. To experience it as often as possible, I focus on things that are beautiful and interesting. And finally I try quite concretely to maintain confidence in life. Do you see the group in this photo?

They're students cooking together.

They're aspiring chemists from all different regions of the Middle East: Syrians, Israelis, Palestinians, Saudis, Iranians. Young men and women. We recently brought them together in Jordan for a conference. While bombs are going off in their native countries, they're trying to understand molecular bonding, nine hours a day. The work is hard, but the shared toil makes them all the more exuberant in the evening—and binds them together. Molecules are only the pretext to create human bonds. Experiments like that give me hope.

Do You Remember?

Neurobiologist

Hannah Monyer

on memory

◆

OUR BRAINS ARE TIME MACHINES. All we have to do is think of the past, and we're instantly immersed in it. But memory consists of fragments; somehow images, smells, feelings from the past are reassembled into a whole. The neurobiologist Hannah Monyer investigates how that happens. Her Heidelberg lab enjoys international renown, and for her work she has received the highest German award for scientists, the Leibniz Prize.

Sometimes memory transports us into a world that seems so familiar to us and yet so strangely remote, as if we were reliving not our own past but a dream. That was how I felt the first time I heard Monyer's voice. She spoke in a distinctive singsong and rolled her "*r*'s"—the speech melody of my long-deceased grandparents. Like them, Monyer is a member of a Romanian-German ethnic group, the "Transylvanian Saxons," who, when talking among themselves, still use a Middle High German dialect like that of their ancestors in the years around 1200, when they first settled in Transylvania. There Monyer was born in 1957. In 1976, she left for Heidelberg, where she studied medicine and in 1994 became professor of clinical neurobiology.

The scientist escorted me into a sparsely furnished seminar room; she put a cookie tin on the table. In it were crumbly confections made of dough and filled with apricot jam, like the ones I used to snack on in my grandmother's kitchen.

◆

Professor Monyer, if you had to lose all your memories but one, which would it be?

Just one? Well, all right: I was lying in the high grass under a huge apple tree in my grandparents' garden. I could hear the buzzing of the bees my grandfather kept there. I had just passed the entrance exam for one of the two best high schools in Romania; soon I would be leaving my family's home forever. I was fourteen years old at the time. I sensed that something special awaited me, that life was beginning. And I felt strong. I'd like to hold on forever to that feeling of calm before setting off.

What did you expect from your life?

Actually, I always knew that I wanted to understand the brain. One day I came home from elementary school and exclaimed, "Mommy, my brain is telling me that I feel pain!" In class we'd just covered the nerve pathways in the spinal cord, and I had grasped that all my sensations were nothing but signals from the brain. At the age of ten, I was immensely fascinated by that. I knew that to learn more I had to study medicine.

Many neuroscientists are driven by the desire to understand more about themselves. You too?

To some extent. But that longing was more important when I was working in psychiatry. The abnormal attracts us so much because we ultimately recognize ourselves in it. I was happy as a doctor. I got involved in research purely by chance, when I ended up in a neurobiology lab on a research grant in the United States. At that point, I knew: This is it. There's this wonderful Greek depiction of Kairos. . . .

The god of opportunity. He's bald except for a lock of hair on his forehead. You have to grasp him by it, or else he'll pass you by.

We believe that we're in control of our lives. But in reality we can only seize opportunities.

Your current research is focused on interneurons. What are they?

Something like metronomes in the brain. Thousands, sometimes millions, of neurons participate in each of your experiences. Their activity has to be coordinated. Special brain cells known as interneurons are responsible for that. Each of them is connected via about 15,000 synapses with hundreds of cells and makes sure that the right thing happens at the right time.

It's only because this interplay works that we can remember scenes from our life.

Yes. When we switched off these cells in mice, their memory was impaired. Apparently, interneurons are crucial to generating synchronous activity, which is required in order to form coherent representations.

They're the conductors in the orchestra of memory.

You can put it that way.

What actually causes us to remember?

Many voluntary and involuntary stimuli can bring up memories. For many people, odors are the strongest trigger for scenes from the past. I can't walk by a freshly mowed meadow without thinking of the village of my childhood. And floor wax has an even more powerful effect—in our home the floors were scrubbed every Sunday.

Why odors in particular?

Because the olfactory sense is most closely connected with emotional systems in the brain. There is, so to speak, a direct nerve pathway from the nose to the amygdala, a brain structure strongly involved in emotional expression. And for good reason: We decide what food to eat based above all on smell. Most animals also make the choice of a mate with the nose. For us humans, how well we can smell someone might be somewhat less important, but we bear this inheritance within us. The olfactory bulb is one of the few brain areas in which new gray cells are produced throughout life. And as we recently discovered, newly formed interneurons migrate from the olfactory bulb into other regions of the brain—as if on ant trails.

A fountain of youth in the brain?

So it appears. These processes must have something to do with the power of memory as well. But we don't yet know the details.

You manipulate the genes of mice in order to alter targeted brain structures.

That's the first step. The second is to investigate the brain activity and the behavior of the animals. That way we gain insight into what the particular systems in the brain are responsible for.

But what makes you assume that these animals remember the same way we do? Maybe the information stored in their heads isn't assembled into a film of the past the way our memories are.

We can only speculate about that—we can't ask the mouse, of course, how it experiences its memories.

It's not even clear whether the animals feel emotions.

No. Still, the fundamental processes are the same for mice and humans: The same molecules and the same networks in the brain are in play. But we have something additional that, when it comes to mice, we know nothing about—subjective experience.

Do you think that we will one day be capable of explaining subjective experience in terms of the effects of molecules and brain structures?

Yes, though I won't live long enough to see this. But the arts, especially literature, offer another, no less valuable, avenue for understanding our mind.

You didn't write your dissertation on, say, molecular biology, but rather on jealousy in Marcel Proust's *In Search of Lost Time*, probably the most significant literary exploration of the phenomenon of memory.

Well, I was interested at the time in how artists viewed illness. To this day, I can't go to an art gallery without diagnosing the figures in the paintings.

In the first chapter of *In Search of Lost Time* there's a famous passage in which the narrator drinks a cup of lime-blossom tea, as he used to in his childhood. The aroma of the tea sets in motion a chain of memories in which his whole childhood comes back to life. Proust writes of a "powerful joy" at that moment. I've always wondered about that— because the childhood that resurfaces for the narrator was anything but joyful.

I don't think that Proust means the joy of remembering. Rather, he's talking about the joy of having discovered a principle: The chance reencounter with the lime-blossom tea has revealed to him how memory itself works. He compares it with origami, the Japanese art of paper folding. When the pieces of paper are steeped in water, they unfold, and whole landscapes emerge from them: A world opens up. It's the same way you feel in the lab when, also usually by chance, you make an important discovery. An incredible joy takes hold of you—it's like falling in love. Experiences like that are granted to us only a few times in our whole life.

A novelist and a scientist might experience the euphoria of discovery in similar ways, but the paths they take to get there are very different.

You perform experiments that have to be repeatable by anyone. Your individuality is kept out of it. Proust, however, describes an inner world from his entirely personal point of view.

Yes, but even though his story is subjective, he expresses a truth that applies to all of us. We read him and enjoy him because he strikes a chord in us. No one would be interested in an author if what he wrote didn't seem as comprehensible to us as a scientific experiment.

Proust discovered that a tiny trigger sufficed to bring back to life a past that we deemed lost. He mentions smell and taste perceptions as the key catalysts; for me, it's often music.

Of course. Sounds too move us deeply. Soon after I had come to Germany from Romania, I happened to hear a symphony by Enescu. . . .

The Romanian composer who is rarely played here.

I immediately burst into tears—even though I almost never cried as a child. It's only memories that arouse this grief.

Because you think of the home you left behind?

Yes. And the older you get, the more preoccupied you become with your memories. In your youth, you're almost indifferent to them. But now, at fifty, I've reached an age when you also look back.

You left Romania at the age of seventeen. Were you oppressed by Ceauşescu's regime?

No, I was lucky. I went to an excellent school, had fantastic teachers. But I saw no future for myself. I wouldn't even have been able to specialize in a particular field of medicine. Attending conferences abroad was completely out of reach.

You must have already been very ambitious back then.

I've always believed in my abilities. And I knew for sure that I wanted to go to Heidelberg. I imagined that it must be something like a German Oxford—a center of intellectual activity. So I applied for a passport for a vacation trip to Germany. I claimed that I wanted to see the country from which my ancestors had emigrated seven hundred years ago. I didn't even tell my parents that I planned to stay there. Only my brother knew.

Did you find your actions cruel to your family—and to yourself?

It was a purely rational decision. I thought my parents would understand. After a few years, I would be able to visit them again with a German passport.

Didn't you ever feel homesick?

Yes, all the time. But it got really bad when I attended my high school reunion thirteen years ago, because at that point, I realized: The country I knew no longer exists.

The Transylvanian Saxons who endured Ceaușescu's regime flocked to Germany after the fall of the Iron Curtain. Only a few old people stayed behind.

On that trip, I visited one big church after another. I'm not religious, but I used to love singing in church. And for eight centuries, those churches were at the heart of our culture. Now they're empty. Of course, other, greater cultures have collapsed. But there's still something special about this minority that held its ground there for so long. For example, the Transylvanian Saxons introduced compulsory school attendance of seven years for everyone at a time when no one in Europe had thought of such a thing yet.

My father is from Transylvania. A few years ago, I visited the city of Schäßburg, now Sighișoara. There I grasped how much education meant to my ancestors' people. The high school is enthroned on a hill high above the city, like a temple. And so that you can get there without getting wet, they've built a covered staircase.

As a young girl, I would spend all my pocket money on books. But I was lucky to live there at a time when the German-speaking minority was no longer sealed off from the Romanians. The compulsiveness of the Transylvanian Saxons could also drive you crazy.

Lightness, gregariousness—I learned all that from the Romanians. Those were my best years. That coexistence of cultures doesn't exist anymore either. . . . I'm sorry. . . .

What's bringing tears to your eyes?

In the end, all of us who left have failed in a certain sense. Faced with the decision again today and knowing that things could actually change, many would definitely decide differently.

Transylvania as you knew it exists only in memory.

Yes.

For me, that region was never anything but memory—secondhand memory. My family members were always talking about "home." They idealized their land of origin. But when I wanted to see it with my own eyes, my father refused to come with us. He said he didn't want to destroy his memories.

I can relate to that. My mother, who is now in Germany, stopped going back several years ago too.

Is it possible to destroy memories?

Not really. One can suppress them, but that's a different matter. Nevertheless, encountering a reality that is completely different from your memories can be very painful.

I was in my mid-thirties when I first saw the region my ancestors come from. And you know what? At first glance, I actually found the fantasy world I had heard so many stories about. We hiked in huge beech forests and among blooming fruit trees, walked through medieval cities and villages in which geese roamed the streets. Maybe memories change your view of the world even when they're not your own.

The villages are beautiful. But you're right. Our perception of reality is always tinged by memory—for all new signals that enter the brain are immediately compared with the already-stored information. That's one reason that the older we get, the more trouble we have getting excited about things. When were you last enthralled by a movie? Compared with the past, movies have generally gotten better. It's just that we've seen such an incredible number of them.

Memories, even good ones, are also a burden.

Certainly.

Memory and experience are, from a physiological perspective, almost the same thing—for something like an organ named "memory" doesn't exist in the brain. The same networks of neurons that process sense impressions in the present also store the impressions from the past.

That's true. But we don't yet understand exactly how those long-term memories are constituted. The mechanisms are different from when you just retain a telephone number for a few minutes in your short-term memory; in that case, the connections between particular neurons are essentially reinforced. When we remember something for the rest of our life, the formation of new neurons might also play an important role.

But every memory, without a doubt, alters the structure of our brain. This insight led the American cognitive psychologist Daniel Schacter to claim that our memories make us who we are. Do you agree?

I definitely see it that way. That's why my preferred gifts to give are trips. The shared memories are very special. The fact that our whole life story is inscribed in the brain might also explain the profound fear many people have of neuroscience.

Because they're afraid of losing their mysteries?

Because they're afraid that everything that makes up their own life could end up under external control. I don't believe that. Even if we understood every detail in a brain, we wouldn't comprehend the whole. And because everyone has a completely unique life story, the networks in the brain are wired differently in each person. Human beings will remain mysterious.

In the science fiction movie *Total Recall*, Arnold Schwarzenegger plays a man who has artificial memories—recollections of experiences

he never actually had—implanted in his head. Can you imagine something like that being possible one day?

Absolutely. Another question, of course, is whether it would be
desirable. But that possibility is so remote that I honestly don't
worry about it. It's enough for me to understand the brain the way
it is.

Let's take a manipulation of the brain that's closer to home: According to a poll of scientists by the renowned journal *Nature*, 25 percent
of respondents take drugs with varying frequency solely with the goal
of enhancing their mental performance. Would you dope your brain
with such substances?

No, because I don't know what that would achieve. Even very sophisticated manipulations almost never manage to really boost
mental retention or other capacities of the brain. We can genetically manipulate our mice as much as we want—ultimately we
always get animals with reduced abilities compared to those that
nature has to offer. It's not so easy to compete with millions of
years of natural selection. Would you do something like that?

I once spent an afternoon with the American chemist Alexander Shul-
gin. He was the first to synthesize the drug Ecstasy and still does so.
Afterward, I was annoyed with myself that I didn't ask him for one or
two pills. I'd definitely have gotten prime goods.

Maybe I'm actually afraid of the loss of control. But if I were going
to experiment with something, I'd be much more interested in
trying meditation. I'm a restless person. When I walk through the
woods in the morning, I resolve to take in the trees for just five
minutes and not think about work for once. I don't succeed.

Are there days when you curse your lab?

Anyone who works with this level of intensity has to give up a lot
of things. Other people have richer social lives than I do, and they
have families. I never had kids because I didn't see where there
was room for them in my life. That was the biggest sacrifice for
me. Looking back, I think it would have worked. That's why I

encourage young female colleagues in my lab to just go for it. But they have it easier; a lot has changed in the past twenty years.

Many people who devote themselves so fully to something hope for glory.

It's an illusion. In reality, the individual is irrelevant to science. The experiments don't bear any personal stamp. And if I didn't do them, someone else would a few months later. Forty thousand scientists attended the last annual conference of the American Society for Neuroscience! In science, the individual doesn't count. I was less interchangeable as a hospital doctor than I am as a researcher.

So, then, what does research give you?

What does it give me? The most wonderful moments in my life. When a colleague informs me of results for which I've been waiting months, sometimes years, and suddenly, as in a puzzle, the connections become apparent, I can completely forget myself. You pause for a moment, think about only this one thing—and not, as usual, about the next thing—and feel nothing but a deep connectedness with the world. Not even memory matters anymore. That's an almost mystical experience—to be, for a few seconds, completely in the present.

The Others in Our Heads

Neuroscientist

Vittorio Gallese

on empathy

◆

FOOD LOVERS ASSOCIATE PARMA WITH PROSCIUTTO and cheese;
opera aficionados think of the home of Verdi. Few people know
that, in this city at the edge of the Po Plain, one of the most significant discoveries in contemporary neuroscience was made. In
the 1990s, a group of young doctors who spent their free time
experimenting in the physiology lab at the University of Parma
stumbled on very special neurons—gray cells to which we owe our
ability to mimic, to empathize, and probably even to speak. The
discovery was celebrated throughout the world. But the heroes
who made it did not take the route, common in the globalized
research community, of assuming academic posts on both sides
of the Atlantic. Instead, they returned to Parma to continue their
research together.

Vittorio Gallese is one of them. Born in 1959 in Parma, he can
delve into the details of a goose liver dish at a restaurant with the
same enthusiasm with which he usually talks about the amazing
abilities of the motor cortex. And over the door of his office—the
place where, elsewhere in Italy, a crucifix would hang—Verdi's portrait keeps watch.

◆

Professor Gallese, you made the most important discovery of your life when you were a doctor at a prison.

At that time, I had completed my service as a medical officer in the air force and wanted to do research. There was no position available for me at the university. But there was an opening at the prison. So I did unpaid work in the lab during the day, while I earned my money nights and weekends in the jail. That went on for five years. In 1992, I finally got a job—at the University of Tokyo.

When did you sleep?

Not often. But the experience in the prison was very enriching for me from a human perspective.

Did you empathize with your patients? You must have known why they were in jail.

I tried to find out as little as possible about their criminal records. As a doctor, I was there to heal, not to judge. But of course that was impossible most of the time, because the crimes were in their files and in the local press. Interestingly, I still felt for the inmates—even serial killers and men who had dissolved their victims in acid. The guards were always asking me, "Why do you bother helping someone like that?"

Yes, why?

If I'd only read about the criminals in the news, I too probably would have felt nothing but repugnance for the killers. But those men stood before me in the flesh, talked about their wives, had a personal history as I did. They weren't completely alien beings. And, not least of all, we shared an environment. Seven doors closed behind me on the way from the street to my office; I knew what it was like to be cut off from the outside world. Because I ultimately lived with them, it wasn't hard for me to put myself in my patients' shoes. Empathy isn't just there; it arises from a situation. We're now beginning to investigate such effects systematically: How does empathy change with the environment in which people encounter each other? And in what ways does it depend on our genetic makeup and personal history?

But it wasn't the prison that prompted you to devote yourself to these questions as a scientist.

Not at all. My interest was much more basic: At the university, we initially just wanted to better understand how the cerebrum gives the muscles instructions. At the time, we didn't yet suspect that we were also on the trail of empathy.

What you discovered was a mechanism by which the brain can, as it were, read the minds and emotions of other people. Some of your colleagues have declared that your discovery was as significant as the decoding of DNA. How do you feel after a success like that?

Now mirror neurons have even been mentioned on the TV show *House*. But to be honest, I don't think about it that much. And I don't know whether discoveries in fields as different as genetics and neuroscience can be compared so easily. Finally, I didn't discover mirror neurons on my own. We were and are a group of scientific equals, many of whom, by the way, were at the time working without pay. My now-wife, Maria Alessandra Umiltà, has been part of the group since 1997. We did know from the beginning that our discovery was a very important one. And what continues to drive all of us in this work is the possibility that the scope of our discovery might be even greater than we can currently gauge.

On the other hand, it's said that you never would have become a famous neuroscientist if you hadn't swiped peanuts from one of your monkeys back then.

Well, it really was a chance discovery: We were recording electrical signals from gray cells that control the monkeys' movements. Whenever the animals reached for the food, those neurons fired, and we would hear a crackle in our measuring instruments. But when I myself at one point extended my arm toward the nuts, the crackle occurred too—as if the monkey had moved. But it was only watching quietly. At first, of course, we thought there had been a mistake. After a while we realized that the monkey's brain actually behaves as if it were putting itself in our shoes. When an animal observes another's movements, the observer's neurons mirror the other's behavior. That's why we called them mirror neurons.

The same thing is happening in my head right now as you reach for your coffee cup: Part of my brain is resonating with yours, so to speak.

That's right. Just recently a colleague from Los Angeles reported on mirror neurons in people. Until then we had only indirect evidence that they exist.

Which finally casts aside any remaining doubts about the reasons for the success of sports viewing. Millions of people on their living room sofas not only watch Michael Ballack play soccer—they *are* Ballack!

At least until he disappears from the screen. Only the resonance isn't equally strong for everyone. For an amateur soccer player who is himself skilled in the movements he is seeing, the mirror neurons are activated more strongly than for someone who never leaves the sofa.

If my brain so closely tracks the movements of others, why don't I actually execute them? What keeps me in my seat while Ballack is dribbling?

The chain of command in your head is blocked at a later level. But often this inhibition is loosened—then people spontaneously mirror the person in front of them. Soccer fans jump up when they see a fellow fan do the same in the stadium.

It's well known that laughing and yawning are contagious. . . .

And for people who suffer from echopraxia, the inhibition doesn't work at all. They compulsively imitate everything the person in front of them is doing. A French neurologist described how he was walking with a patient like that along the railing of a hospital balcony, opened his pants, and peed off the balcony. The poor man couldn't help doing the same.

Apparently, mirror neurons don't just enable us to put ourselves in others' shoes. When the right impulses are triggered in my brain by mere watching, then that mechanism seems almost designed for us to adopt new behavior—by copying it from others.

Our extra-long childhood, compared with that of other creatures, is conducive to that. As a result, human beings have far more mirror neurons than any other animal. A chimp has to watch for five years before it can crack a nut on its own by using a stone as a hammer and another as an anvil. A small child can learn that in a few minutes.

In order to improve my rowing technique, maybe I should practice less and just watch the German rowing champion more.

Even better, you'll increase your muscle strength by just observing another person's movements. A series of experiments have recently shown that; one of them used Japanese weight lifters. It's probably because the brain learns to control the contraction of the muscles more effectively.

So, then, why keep paying for a gym membership?

Because there you'll build your biceps an additional 20 percent. We're now beginning to investigate how we can take advantage of resonance phenomena like that for medical purposes. After a stroke, for example, it might accelerate the rehabilitation considerably if the patients watch videos of the correct movements.

Amazingly, not only the movements of the person in front of you are mirrored in your brain, but also their intentions.

We know that from experiments in which we performed actions out of the monkeys' sight and the animals only heard us. Their mirror neurons fired anyway. And even more than that, their mirror neurons can actually tell why I pursue a certain intention. Depending on whether I reach for the cup to drink it or to clear the table, different neurons are activated in their brains. We've shown that in experiments too.

We have neurons that read others' minds. How can individual gray cells be so smart?

They receive information from many other centers in the brain—in the case of the coffee cup, for example, about whether it's already empty or not. And it's not as if you were born with mirror neurons for behavior at the breakfast table. The system has learned what people do in their environment.

And then we grasp it without having to think about it.

Exactly. We experience other peoples' intentions as if they were our own. Psychologists were fundamentally mistaken in that respect. According to prevailing opinion, I have to understand myself before I can comprehend your intentions. But that's not true: In most cases, I don't need any theory at all about mental states, whether my own or yours—because the mechanism of the mirror neurons offers us direct access to the internal world of other people. Only autistic people are forced to take the indirect route of always first thinking about the other person.

Why do you mention autistic people in particular?

Ask them yourself! They'll answer that they can't empathize. That's why they always have to consider what might be going on in the minds of people they're with—that's difficult and goes awry all too often. We have evidence that autistic people's mirror mechanism is impaired. When healthy children watch you eat strawberries, their mouth muscles are automatically activated each time the fruit nears your face. For autistic children, that's not the case. As a result, those children have unusual difficulties learning sequences of movements. Actions and feelings are connected.

Can empathy be trained?

One promising approach might lie in improving bodily awareness. We're currently trying to find out whether people with autistic disorders can be helped in that way: Dance, acting, and musical performance could contribute to the improvement of motor abilities and, with them, the capacity for empathy. We've also just begun experiments to learn what happens in the brains of autistic people when it comes to the sense of touch.

My brain reenacts not only other peoples' movements and intentions, but also their sensations. When I see someone being caressed, this activates the centers responsible for touch in my brain, which are also equipped with mirror neurons. And the parts of the brain that generate the sensation of pain react exactly the same way. The phrase "I feel your pain" should be taken literally.

Not entirely. Your systems for pain switch on when you see me, say, in a dentist's chair, and probably you screw up your face when the drill approaches my mouth. But your brain doesn't receive any pain signals from your own body. That's how it concludes that it's my problem and not yours—the sensation is blunted.

That's probably the difference between empathy and sympathy: I might slip into your skin, so to speak, but, paradoxically, I still don't have to share all your feelings.

The key is that you put yourself in my shoes intuitively as opposed to intellectually—even if the actual feeling itself is barely triggered. Actually experiencing another's feelings doesn't happen until the next level, when you might sympathize with me, when you, say, "feel my pain." But that occurs much less often.

If empathy is an automatic mechanism of the brain, there must be a high threshold for sympathy. Otherwise, we ourselves would immediately feel all the suffering we observed. The endless cruelties of history would never have taken place. Then again, surgeons and dentists never would have existed either.

You can actually separate empathy and sympathy from each other completely. Just think of a sadist, who takes pleasure in his victim's pain precisely because he can empathize with him. Empathy doesn't at all guarantee altruism.

What determines whether empathy turns into sympathy?

That's the key question. We still know very little about it.

Mothers often say that they experience their children's pain as their own.

Colleagues of mine in Rome have recently shown that the systems for pain in the brains of nursing mothers have a particularly strong response to videos of screaming infants. But that activity is even higher when you show them their own baby. And only in that case do the areas of the mother's brain that control movements become active. Even before they realize it, women are clearly preparing to help.

Sympathy and altruism are apparently interdependent in the brain as well.

That seems to be the case. But we still know really little about those connections too. I personally believe that altruistic behavior isn't innate to us—unlike our capacity for empathy.

New, well-controlled experiments show that children all over the world start sharing with others at the age of four.

But even if they behave that way on every continent, that doesn't prove a genetic cause. Maybe people in all cultures are taught to be kind to one another. That might be supported by the fact that children don't begin to share until the age of four or later—that's how long it takes to overcome their natural egoism. In any case, mirror neurons alone don't make us better people. In my view, our discovery contributes in an entirely different way to improving our understanding of why we act morally: All of us have a sort of apparatus in our head by means of which certain customs can spread very easily among people—because we simply copy them.

> In fact, the human sense of morality can be completely separated from empathy. Often we act morally even when we don't feel especially close to another person at all. A lawyer, say, might take on the case of a murderer as a court-appointed defense attorney, even though he can't identify with this person and wouldn't even want to. I doubt, however, whether something like morality could have evolved in the first place if humans had not been at least inherently capable of emotionally taking the perspective of others.

Probably not. Instead of intuitively empathizing with other people, you can make an intellectual effort to understand them. But what you experience in that intellectual process has an entirely different quality from empathy. For example, I didn't become a father until late in life, at the age of forty-five. From my friends' accounts, I knew, of course, what it might be like to have children. Theoretically, I had all the information I needed to understand the experience. But I realized what it really means only when my daughter was born, when I could hold my own baby in my arms, when my experience was physical. Since then, I can relate to other parents much better. As long as you only know someone else's situation, you'll misunderstand a lot more than you would if you could empathize with them. That's probably why we instinctively gravitate toward people who find it easy to put themselves in our shoes.

> Are there people who are by nature more empathetic than others?

Certainly. It probably has to do with how much you mirror others' facial expressions. The brain constructs emotion from the movements of the facial muscles. When the mouth and the corners of the eyes display a real smile, our mood is elevated; if we make a sad face, however, it is lowered. In experiments, it turned out that people who unconsciously take on someone else's expression to a greater degree are at the same time more empathetic.

> Those are the people who burst into tears with Vivien Leigh at the end of *Gone with the Wind*.

Yes.

> Strangely, we seek out experiences like that. We want to be moved
> by a movie or a play. Why?

Possibly it does us good to occasionally see others suffer. Are you
familiar with the French philosopher of religion René Girard? He
argues that actors on stage are symbolically sacrificed. In that way,
society discharges its ever-present propensity for violence without
doing harm. I think that theory has a lot going for it.

> You mean that Tristan and Isolde die for love so that I can better bear
> the unavoidable trouble in my marriage?

A fantastic opera! Even though it's not by Verdi. I saw it in Tokyo,
with René Kollo in the title role.

> But at least as powerful as the discharge of aggression is the oppo-
> site effect, the fueling of aggression. Observed violence can be con-
> tagious; some scientists claim that young people in particular imitate
> the crimes they see in movies and video games.

I'm skeptical about that, because I don't know of a single study
that really proves a connection between our innate ability to imitate, media violence, and brutality in real life. But even if violent
video games and movies don't provoke aggressive acts, they can
desensitize people to the sight of cruelty, so that violence becomes
banal and eventually even appears to be an acceptable solution
to conflicts. But it should be emphasized here that never before
in history have so many people coexisted so peacefully as in the
present. Most epochs were far more brutal, without any video
games. In terms of the impact of technology and media, there are
developments that I find much more troubling.

> Such as?

The encroachment of the virtual world. We communicate more
and more by telephone and computer; communities in which
people encounter one another in person are increasingly dis-
solving. But we know from our experiments that it's not without
consequences for our capacity for empathy whether we see

another person only on a screen or face-to-face. That's why a theater experience is more powerful than going to the movies. And if you communicate with your conversation partners only by e-mail or, like many young people, in electronic chat rooms, your image of them dissolves completely.

We disembody the people we associate with.

Yes. And that must have profound effects on our social and cognitive abilities. We just don't yet know what they are. In any case, our social skills evolved for direct encounters, not virtual ones.

But you yourself benefit from the fact that you can now work with colleagues in Japan via the Internet almost as if they were sitting in front of you. Regardless, the development of technology can hardly be stopped.

And I wouldn't want it to be. Still, our conversation would definitely take a completely different course if we were talking on the telephone instead of sitting across from each other here in Parma. If electronic communication spreads more and more widely, we'll probably have to find completely new forms of interaction. Video telephones would be one small step.

Without seeing an image of another person's body, you can't truly empathize at an intuitive level.

No. You can only try to understand others at a theoretical level, much like autistic people do. But that way is far more complicated, and you'll make a lot of mistakes. That's why we seek the company of people who relate to us without long explanations; all they have to do is look at us.

In that way, people who don't otherwise share many interests can have the best friendships. . . .

Because other commonalities are far more critical. Friedrich Nietzsche was right when he wrote, "To understand another person, that is, to produce his feeling in ourselves. . . ." If two people can't do that, even marriages between partners who seem as if they were made for each other will fail.

I don't know any other neuroscientist who cites thinkers like Nietzsche, Husserl, or even Heidegger as enthusiastically as you do. What do you get out of philosophy?

Many of my colleagues see philosophical ideas as a sort of embellishment that helps make their research more palatable for the public. That's all right if they're dealing with problems that are very remote from human experience—say, how ion channels transmit electrical signals between neurons. But if they're investigating phenomena like empathy or even consciousness, they can no longer treat philosophy like the decorative cherry on the cake of their research—because what they're studying is inseparable from their personal worldview. It's much harder from the outset to ask the right questions. And for that the philosopher's systematic approach is enormously helpful.

So you're trying to draw on the wealth of 2,500 years of thought about the central questions of human existence?

That would be going too far. It's more that you create your own tailor-made philosophy. You choose the thinkers whose ideas you find particularly stimulating for your own research. For me, those include the phenomenologists around Edmund Husserl. . . .

An Austrian mathematician and philosopher who, around the turn of the twentieth century, explored the corporeality of all objects of knowledge.

And the fertilization of science by philosophy is not a one-way street. Our research leads philosophers to ask new questions too, and so there's a constant cycle of exchange.

To invoke intellectual history so strongly while working in the laboratory seems to me a very European approach. Americans, who are often portrayed as paragons of science, are much more pragmatic in that respect.

Yes, they focus their interest more intensely on immediate solutions to problems. And they have no choice in that, because there's far more competition in their universities. They have to produce results—and can't afford to take great risks or think about far-reaching theoretical consequences of their research. It seems to me that our cultural heritage will help us Europeans continue to make significant contributions to science.

What can philosophy learn from neuroscience about empathy?

That the human mind is inconceivable without the body.

You're contradicting a view that seems to many people almost self-evident since the triumph of computers: the notion that the human mind functions like a computer, that your character and your memories are only a huge collection of data. If that were the case, you could at least theoretically transmit all the information that makes up your personality from your brain onto a supercomputer and run the program there. In a way, you would then live on in a silicon chip.

Complete nonsense. As we now know, all our thoughts and feelings are dependent on the fact that we observe the bodies of other people, that we touch and manipulate things. And there's increasing evidence that we owe even the ability to speak to such motor skills. Our mind exists only in the corporeal world.

And is therefore hopelessly mortal.

Definitely. But by pursuing far-reaching goals, we make that prospect bearable for ourselves. When the emperor of China ordered the construction of the Great Wall, he knew that he would never live to see its completion. Still, he put all his energy into that undertaking. Have you by any chance ever read the poet Ugo Foscolo? We read him in school. In one of his works he describes a visit to the tombs of Santa Croce in Florence, where Michelangelo, Galileo, Rossini, and many other significant men are buried. But in the brains of later generations they live on. That too is a sort of resonance. Thanks to them, humanity was able to create culture and liberate itself from the constraints of biological evolution. And even by passing on just a little bit of knowledge, every human being makes a contribution to that culture, which transcends mortality. But maybe you have to be Italian, stumbling on traces of the past with every step you take, to feel that way.

The Laws of Devotion

Animal behavior researcher

Raghavendra Gadagkar

on altruism

◆

CAN WASPS, OF ALL THINGS, contribute to saving the world? I first encountered the entomologist Raghavendra Gadagkar at a conference in the Swiss Alps, where several Nobel laureates as well as high-ranking politicians and a former UN High Commissioner for Refugees explored new ways to solve conflicts and strengthen global understanding. There Gadagkar gave a brilliant talk on a primitive wasp named *Ropalidia marginata*, which populates his native South India—an insect that, to an untrained eye, is scarcely distinguishable from the creatures that terrorize us in late summer when we eat dessert outside, but has a far more interesting social life.

With his research on the nature of conflict and cooperation, Gadagkar has become one of the leading animal behavior researchers in the world—and a corrective to the myth that outstanding scientific work is necessarily conducted only in places like Harvard, Oxford, or possibly Munich. Born in 1953 in India, Gadagkar studied biology at the Indian Institute of Science in Bangalore and became professor of ecology there. On the side, he founded in Bangalore an institute dedicated to bridging the divide between the humanities and the sciences.

For this conversation, we met at the Berlin Wissenschaftskolleg (Institute for Advanced Study), where Gadagkar is a nonresident permanent fellow. It was one of those dark-gray German winter mornings that demand an almost religious belief that the sun will someday return to the sky. Gadagkar, however, who had just arrived from the heat of Bangalore, found the cold drizzle "refreshing."

◆

Professor Gadagkar, you live with the objects of your research. It is said that your whole house is full of wasp nests. How does your family feel about that?

We've even named the house "Ropalidia" after my wasps. The insects are simply part of my life. In the past my wife took care of them when I traveled, and now my students do. The first time our son was stung, he was delighted—he felt like a grown man.

And how often have you yourself been stung?

Very often. Nowadays somewhat less frequently than in the past, though, because my students do most of the work with the insects. But as long as they stay focused on the wasps, nothing happens to them. Wasps sting only when the students' minds are elsewhere— for example, if they make a jerky movement.

Every single wasp in your colonies is marked with tiny colored dots. Do you anesthetize the insects in order to paint on them?

Oh, no. You just have to wait patiently near the nest until the insect is busy with something and is looking away, and then you can paint its thorax with a toothpick.

How did you become interested in studying wasps?

It started as a hobby. When I was an undergraduate, our college buildings were full of wasps. They really nested everywhere! So I began to observe the insects. I read a lot about them and ultimately even published a few research papers. But I did my PhD in molecular biology. To really advance that career, however, I would have had to leave India and go to the United States or Western Europe. I didn't want to do that; I would have missed our culture too much. So I decided to make my hobby into an occupation and take up animal behavior research. I never regretted it.

What do you find so exciting about these insects?

I look at them the way an anthropologist looks at a foreign culture. We have our own society in mind and have difficulty comprehending that a different society can be completely different—and sometimes surprisingly similar. The insects hold up a mirror to us.

Wasps seem to me very remote from us humans.

You know, if you watch them long enough, you realize that wasps have personalities. Each of them reacts differently, has its own strengths and weaknesses—and even seems to be aware of them. At least, each individual seeks a place in the society that suits its abilities. To observe that is fascinating: There's competition and cooperation as in a human society.

"There is no such thing as 'society,'" Margaret Thatcher, the former British prime minister, once declared. There are only individuals, she argued, who have to fend for themselves.

She was mistaken. The Iron Lady should have taken a look at a wasp colony.

Thatcher could have invoked Charles Darwin. He claimed that each creature competes with every other creature for resources and the best reproductive chances.

Yes, but Darwin himself was aware of a significant paradox in his framework. Something like a bee sting couldn't exist according to Darwin's theory of evolution. The fact that a bee dies when it stings but does it anyway really confused him.

The bee commits suicide for its family, so that at least the genes it shares with its sisters live on.

That's the explanation, as we now know: The theory of evolution applies, but it has to be extended to genetic relatives. That's why the related workers among my wasps also abstain from having offspring of their own. Instead, they all provide for one among them, the queen, who is the only one to reproduce.

How is it decided who becomes queen?

Among honey bees or vespine wasps, the queen is born the queen. My wasps are much more interesting: Theoretically, any female can found her own colony as queen. But only the particularly fertile insects do so. The rest submit to her and take care of the queen's offspring. Because the queen is usually related to everyone in the nest, their own genes are passed on in that way too—and even more reliably than if each worker had offspring herself. Somehow, the insects seem to know that. But how? We have spent the last twenty-five years trying to answer that question.

A perfect socialist society: Even reproduction is collectivized. Members of the 1968 student movement here in Germany would have been delighted with your wasps.

Or, then again, maybe not—because as soon as an insect has attained the status of queen, she releases chemical substances in the nest, pheromones that suppress reproduction among all the others.

The queen rules by drugging her followers into submission.

You can put it that way. But as soon as her fertility drops, her power is taken away and transferred to another queen. The insects submit to a ruler and cooperate with one another only when it's to their own benefit.

The males in the wasp society haven't even come up yet.

They're raised by the females, lead a brief nomadic life, mate, and die. They never work for the community. We wondered why not. Our experiments showed that the male wasps, the drones, could contribute very well to the care of the larvae as long as they're in the nest. But they don't, because the females do the job much better. Basically, the males are superfluous, as far as work is concerned.

That's actually the case everywhere in nature, isn't it?

Yes. Among wasps as well as bees and ants, it's especially clear: Males aren't necessarily even indispensable to reproduction. The queens' sons hatch from unfertilized eggs. Only to have daughters do the queens have to mate. Then they receive the sperm in a pouch in their body where it stays fresh and can be used as needed. So the queen has complete control over the sex of her offspring!

Couldn't the queens also produce female offspring from unfertilized eggs? Then the insects could spare themselves the rearing of males altogether. Why is there sex at all?

That's one of the mysteries of the theory of evolution: A species consisting only of females could reproduce with half the effort—a huge advantage. But they would also be more susceptible to disease-causing agents. Those parasites are curbed by the fact that sexual reproduction constantly reshuffles the genes of their hosts. . . .

Which happens among only some of your wasps, however. All the drones of a colony genetically resemble their mother, the queen. As a result, the offspring are also much more closely related to one another than is the case among the offspring of other animals. That would explain the extreme cooperativeness. All altruistic acts remain in the family.

That's how it was explained for a long time. But we were able to show that things aren't that simple. For example, we introduced one wasp colony into another. As expected, the resident workers tore the foreign queen to pieces. But her daughters, the young workers, were accepted into the colony without any trouble. They even provided the next queen. Now the insects in the nest were no longer all related but worked together anyway.

Which was better for all parties.

That's exactly the point. Far more important than relatedness is whether the benefits of cooperation outweigh the costs. The fact that the significance of relatedness was overestimated for such a long time has to do with a misunderstanding. In a famous formula devised by the English evolutionary biologist William Hamilton in 1964, everything is crystal clear: Cooperation depends, among other things, on the degree of relatedness, but also on the costs and benefits of working together. Only it's terribly difficult to measure costs and benefits in experiments. Relatedness, however, can be determined quite easily with molecular methods. That's why most scientists have simply acted as if that were the sole factor. But when you take a closer look, you realize that even among wasps, not only the genes, but also the environment determines behavior. For example, as long as the circumstances of life are relaxed, a relatively large number of wasps actually go off on their own and build their own nest. If, however, food is scarce and the colony is threatened by enemies, many more insects are willing to devote themselves to the community.

Helpfulness among Londoners has supposedly never been as great as it was during the German bombings. There are similar accounts from New York in the days after September 11. Sociobiology claims that human behavior follows the same rules as that of your insects.

The weighing of the costs and benefits of altruism applies to all species—including us. Of course, the sacrifice we find among wasps, for example, is extreme: Most of the females give up having their own children for the sake of others. We biologists call behavior like that "eusocial." And for a long time we thought that only insects like wasps, bees, and ants were capable of it—because of the unusually close relatedness among the individuals. But we were mistaken. Even such extreme forms of cooperation are more widespread than we thought. These must have evolved in the animal kingdom independently of one another at least a dozen times, including among animals that reproduce in the same way we do. Certain spiders, beetles, and shrimps are eusocial—even a mammal, the African naked mole rat, is. The sociality of these rodents in their underground tunnels is actually quite similar to that of the wasps.

But different from that of human beings, who want to mother their own children.

Still, the principle is the same: The benefit-cost ratio determines the degree of cooperation, whatever the goal animals or people pursue together. It's just that the weighing of costs and benefits is really complicated for *Homo sapiens*. Unlike insects, shrimps, and naked mole rats, we're not only determined by nature, but also profoundly influenced by cultural values.

In our Western culture, there's a commonly held belief that humans are purely selfish by nature, and only upbringing and education make us into moral beings willing to share.

Far more than the morality that we preach, it depends on the conditions under which we live. The lower the cost, the higher the benefit, and the closer the relatedness, the more willing animals as well as people will be to put themselves out for others. If, for example, there's a big age difference between siblings, it doesn't cost the older siblings much effort to do something for the younger ones and achieve a lot. With a narrower gap in age, the ratio is not as favorable, and as a result, those siblings aren't willing to go to as much trouble for each other. In our Indian extended families, you can observe this effect all the time.

But a society isn't a big family. Since every one of us isn't related to everyone else, we're much less willing to share.

Only we've shown that even among wasps, relatedness is only one factor among several that determines cooperativeness. So if you want people to cooperate, you have to set up their environment accordingly. Altruism can't cost too much and has to achieve something.

Can social biology tell us how to live?

No. But it can help us facilitate a particular way of life that we desire. The advantage of studying insect societies is that you can conduct pretty simple experiments with them. For a few years, however, scientists have also been researching the best strategies to promote fairness and cooperation among people. Biologists, psychologists, mathematicians, and economists are coming together to develop something like a science of cooperation. I find that very encouraging.

What have been the results?

It's come to light, for example, that people of all cultures share a powerful aversion to those who take unfair advantage of others. Most of us are willing to punish cheaters even when we ourselves aren't affected by their cheating and even when we have to incur costs to ourselves in order to punish them.

Apparently, that sense of fairness has evolved because it is to the long-term benefit of every individual—even if you have to pay for it in the short term.

It definitely seems to be a very effective strategy. The most selfish people in particular are reliant on the community—they want more from it than they give. Some of my colleagues even believe that the reason we humans cooperate isn't so much that we feel a particular urge to do so but rather that we need to guard against cheaters.

In your culture, the collective plays a much more significant role than in the individualistic West. Does an Indian perspective help us understand how communities work?

Our society is extremely complex. Maybe it was this background that aroused my interest in social insects in the first place. The division of labor and the coexistence of different generations in a wasp's nest are very delicately balanced—much like in our society. And the genetic relationships in the colony are even trickier than in any Indian clan.

Your colleague, the American evolutionary biologist Richard Lewontin, has argued that scientists bring deep-seated prejudices to the issues they investigate. What are yours?

Is it always bad when scientists have prejudices? Strong opinions sometimes give people the energy to proceed undeterred in a particular direction.

But if someone believes, for example, that life on earth arose because seeds from outer space rained down on our planet. . . .

Then let them! If it turned out that they were actually right, we would all be wiser for it. And if their attempts to prove it fell flat, the result would be the same. It's easier, by the way, to go your own way if you don't work at Harvard or Oxford or even in Berlin, but somewhere on the periphery.

Like Bangalore.

Exactly. You're scrutinized less, and you have greater freedom. We need more nonconformists! After I received my doctorate, everyone advised me to continue my research in America, or else I'd never get a good job. At the time, I said I'd accept that consequence. I want an environment in which, on the contrary, we reward people for not conforming.

You're convinced that intelligence and even a certain form of consciousness should be attributed to your wasps?

Intelligence and consciousness can certainly be defined in such a way that insects are automatically excluded. But such linguistic conventions don't interest me, because it's clear that wasps are not merely robots. They learn, and their behavior isn't simply predictable.

What do they learn?

Where their own strengths and weaknesses lie—for example, how successful they are at collecting food. You could say that they know their own personalities.

But does that by itself signify intelligence, even consciousness?

At some rudimentary level, yes. You can't draw a sharp line between creatures endowed with reason here and purely instinct-driven ones over there. It's a continuum.

One of the incidents you've reported from your wasp nests I find particularly strange. It was a sort of rebellion: A group of workers stops working for the queen, designates a leader from their midst, and a few days later founds a new colony—as if they had coordinated with each other.

That was what happened. We don't know how that decision came about. Apparently the wasps have means of communication that remain unknown to us. Somehow they seem to actually "speak" with one another. One possibility is that they exchange information through their saliva—for the workers are incessantly touching, attending to, and feeding one another.

The idea that their language could be in the composition of their saliva I find astonishing enough. But the split that you've described means that the insects must have planned their behavior days in advance. That seems incredible to me.

Yes, we have to draw that conclusion. Animals are capable of considerably more than we currently understand. The fact that we constantly underestimate them has to do with how we've isolated ourselves from them. In modern society, animals have become so alien to us that we've forgotten the ways that many species are superior to us humans in certain areas. We've lost respect for nature.

Do you like to go to the zoo?

No. I don't feel particularly comfortable in the presence of animals in captivity. I'd much rather observe a wasp than a lion or an ape behind bars—even if the wasp wasn't a member of a social species. Even with such apparently unspectacular creatures there's so much to see: How the solitary insect gets some soil, mixes it with its saliva, and builds a nest with the mud; how it then catches a caterpillar, puts it in the mud, and anesthetizes it with its sting; how it fills the nest with eggs and closes it. . . . You could spend your whole life watching and continue to discover new things.

We owe modern biology solely to one man's devotion to watching closely. Charles Darwin must have been a phenomenally close observer. And from what he discovered about the birds and tortoises of the Galapagos Islands, about barnacles, about pigeons he bred himself, he conceived the theory of evolution. He had no tools but his eyes and his mind.

Nowadays science is based on elaborate experiments. Too many scientists rely on expensive equipment, sometimes unnecessarily. I tell my students to spend much more time thinking than using technology. In the race for equipment, we can't keep up with our colleagues in rich countries anyway. We have to make up for this deficit by using our brains. That might be an advantage, because in that way we're more likely to reach a deep understanding.

The theory of evolution has recently turned 150 years old. But many people still haven't accepted it and even attack it. They find Darwin's theory blasphemous and bemoan the absence in it of a divine plan.

That's the case in the West. In India no one has trouble with the theory of evolution. The idea that the world is in a state of constant development, that destruction and creation go hand in hand, that natural history has no goal—all this has been taught in Hindu philosophy from time immemorial.

In your country, almost half of the population still lives at or even below subsistence level. Are your requests for money to study wasps met with great sympathy?

Absolutely. India has a very long tradition of valuing knowledge. And everyone agrees that we can only begin to make a leap forward through the acquisition of knowledge. That is understood even in the poorest villages. But it's important for us scientists to communicate with the public about our work. That's why I try to explain to outsiders as often as I can what fascinates me so much about my research—much to my own benefit, by the way. Often questions from the audience have given me new ideas.

What do you get out of studying nature?

When I watch the wasps feeding their larvae, I become completely calm. Those insects always remind me that I'm part of a community—and how important that community is. Other scientists—say, those who study molecules—can easily forget that. And taking care of animals ultimately teaches you to have a less inflated opinion of your own importance.

You can learn that by taking care of other people too.

Yes. But people almost always give you something back. From a wasp, however, you get nothing at all. In that way, you learn true devotion. That's why we should do whatever we can to ensure that our children develop an intimate relationship to nature.

The Hunger for Fairness

Economist
Ernst Fehr
on morality

◆

WE ASSOCIATE ECONOMISTS WITH TRADE BALANCES, job figures, and forecasts. But the focus of Ernst Fehr's career is the hunger for fairness. Fehr's works are among the most-cited texts of modern economics; he has received major prizes in his field; and it would be hard to name a prestigious university that hasn't wooed him. But Fehr, born in 1956 in Vorarlberg, Austria, has remained true to the University of Zurich, where he has worked since 1994.

It's not only his subject that is unusual among economists, but also his methodology. While economists normally attempt to capture the balances of supply, demand, and prices in mathematical models, Fehr is interested in the human being in the market. In collaboration with neuroscientists, he hopes to unlock the mystery of what drives our decisions. On his desk is a white porcelain head from the nineteenth century. It is marked according to which areas of the head are, based on the theory of the time, the seats of which emotions.

◆

Professor Fehr, when were you last angry about an injustice?

While reading the newspaper this morning. I feel the same way as almost everyone about the discussion of managers' bonuses. When executives with massive incomes who have ruined their companies over the past several years demand and receive a payment of millions, the public gets justifiably upset.

You're an expert on emotions in the economy. That's unusual: Economics is viewed as a science that looks coldly at the world.

That view is incorrect. It's true that classical economics imputes fully rational self-interest to human actions. But certain emotions have nevertheless always played a role in it.

Like greed.

Or fear. Only my colleagues don't call these emotions but "preferences." Just as you might prefer to buy chocolate over sauerkraut, you also have a preference to avoid or seek risks. Emotions play a role in decisions.

If there's a preference for fairness, economics has remained silent about it for quite a long time.

Economics has systematically blocked it out. Of course self-interest is a very important motive. But we first had to gather evidence for why it's a mistake simply to gloss over other impulses such as altruism and the sense of fairness. For a long time, we were met with pitying smiles.

The prevailing ideology was that an economy works best when you just encourage profit seeking. The fact that this went so completely wrong must have been a validation for you.

It was. Unfortunately, though, those who weren't responsible for the disaster now have to foot the bill.

You must have felt a strong need for fairness from early on. You supposedly began studying economics in the hope of contributing to a better fate for developing countries. Did you really believe that?

Yes. I was politicized in the student movement of the late 1970s; at the time, development politics was an important topic.

You belonged to a group called "Roter Börsenkrach," Red Stock Market Crash.

Well, one of our professors convinced me that you have to know well what you criticize. So I studied neoclassical economics; that has benefited me to this day. But at the same time, I always found the framework of my subject too narrow. Instead of Schumpeter, the major Austrian economist, I preferred to read Sigmund Freud.

You also flirted with theology. Did you want to become a priest?

I could imagine it. I come from a Catholic region and had read the Latin American liberation theologians. Leftist Christianity was what first got me interested in politics.

Why did nothing come of theology? Because of women?

No, no. It was more that I realized I wanted a just world here on earth, not a just paradise somewhere else.

If you feel the urge to make the world a better place, people will disparagingly label you a "do-gooder." Does that bother you?

Not if they're just referring to your good intentions. But I defend myself against the charge of naïveté, which I'm so often accused of—because I've thought a great deal about how society can be improved without resorting to naive solutions.

Did you choose the objects of your research with that goal in mind?

Probably. Nowadays, however, I'm less concerned with what constitutes a good society than with investigating how people arrive at judgments about fair und unfair. That's a somewhat different question. We're doing empirical fairness research, so to speak.

What is fairness?

Are you asking me for an abstract definition?

I'd prefer a concrete one.

The matter is most clear-cut when you have a public good from which everyone benefits—like the unemployment insurance program. Now there are always freeloaders, who don't contribute much to the good but take the money. Almost everyone finds that sort of thing unfair.

That might be putting the matter in overly simple terms. When a German man was discovered using his welfare payments to rent a beach house in Florida, the freeloader became known as "Florida Rolf," and was accused of lazing around in the sun at the cost of the general public. This one story lent itself to being exploited as political fodder.

But it's also fodder for science. Because now you can measure what degree of parasitism people find just barely tolerable and what degree they find appalling. We do that with games—a term that somewhat trivializes the experiments. It's all about strategic decisions. The trust game is a popular example: I have ten euros, and you have ten euros. If I give you my money, the researcher will triple it, and then you'll have forty euros. Now I'd like to know how much you're going to give me back of the forty.

I'm playing a bank, so to speak.

Exactly. What is interesting in this game, however, is the fact that I can't force you to pay me back anything at all. But most people do it anyway. Why? Because they have a preference for fairness. . . .

Or they're afraid that if they don't, the other party will never do business with them again.

The trust game even works with players who know that they will never meet again after one round. Traditional economics can't account for that.

If you take the idea seriously that you never know when you might cross paths with someone again, then those games seem quite detached from the real world. How much do they reveal about the far more complicated relations in everyday life?

We don't approach the games in isolation but try to supplement them with systematic studies of real situations. Think of the trust game: In everyday life, too, you're usually honest, even though you don't have to be. You pay the taxi driver, even though it would be easy to just run away; you even leave a tip at a restaurant where you don't plan to eat ever again.

Understood in that way, fairness primarily means behaving faithfully toward your business partners. But that alone doesn't lead to a better world.

Sometimes it can even lead to a worse one. All the evils of corruption would be quickly eliminated if no one felt the urge to make good on a bribe. Or take former German Chancellor Helmut Kohl, who refused to work with investigators when his party was accused of accepting illegal donations because he had given his word not to betray the donors.

A perfect egoist in his place would have better served democracy.

Probably.

The traditional argument would be that norms of fairness were instilled in all of us as children: "A promise is a promise."

Maybe so, but it's interesting that societies developed social norms in the first place and insist on adherence to them.

Because societies wouldn't work otherwise.

They would work, just not nearly as well. Consider the fact that for most of human history, no binding contracts existed at all. So how could something like reliability arise among human beings? We suspect that so-called group selection was involved: There were some societies in which norms mattered more and others in which they mattered less. And the ones in which, for example, people kept promises worked better than the ones in which everyone pursued only their short-term advantage. Some norms might also have been more feasible than others. Societies in which better norms were customary and were also adhered to succeeded. The others collapsed. In that way, certain standards of behavior were gradually established.

If a norm of fairness is so deeply entrenched in us, then to what degree is it innate as opposed to learned?

No one has found altruism genes yet, but we have evidence that something along those lines might actually exist. For example, identical twins behave in strikingly similar ways in games that involve sharing. That's why we're currently performing experiments to search for the genetic bases.

Apparently, it's not only humans who long for fairness. I found new experiments with dogs interesting. They refused to give their paw when they saw that another dog received a biscuit as a reward but they didn't.

We've studied marmosets. Those small primates are even capable of a sort of unselfish behavior: They were willing to work so that a fellow member of their species would get a reward—even when they themselves didn't get anything out of it.

Children also react to unfairness from a very early age. If I bring something back from a trip for my older five-year-old daughter but not for my little one, who's not yet even one year old, there's trouble.

That's not yet a fully developed sense of fairness, however, because small children only worry about whether they themselves are getting shortchanged.

Like dogs.

For humans the concept of fairness begins to expand around the age of five. We discovered this through experiments in which we gave children gummy bears and Smarties. If they wanted, they could share them with other kids. The recipients were not in the room and couldn't pay the giver back later. Under those conditions, three-year-olds weren't very willing to part with their candy. Of the six-year-olds, however, a quarter gave some away. And of the eight-year-olds, 45 percent shared their sweets. A similar willingness to share is found among adults.

What differentiates those who give away some of what they have from the egoists?

We'd like to know that too. Strangely, height seems to play a role—the taller you are, the more selfish you are. And the more dominantly children behave, the less they give. Surprisingly, many only children actually share an unusually large amount.

When we share, the so-called reward system in our brain is activated, which is responsible for feelings of pleasure. Recent studies have demonstrated that. Are altruists happier people?

That might well be. For some people, the reward system already has an unusually strong reaction when they just watch someone else get something.

As if they were happy for the other person.

Yes. Those people then also give away a particularly large amount.

So we could assume that altruism makes us happy. The reverse would also be conceivable, that happy people are intrinsically more selfless.

That's exactly the problem. We don't yet know what is cause and what is effect.

Colleagues of yours here at the university are working with Tibetan scholars who are trying to approach the problem from a Buddhist perspective. The Dalai Lama himself has declared that it's nothing but a particularly wise form of selfishness to help others—because the happiness of having done something good for your fellow human being is more lasting than when I, say, just buy something for myself with my money.

It's important to note, however, that the Dalai Lama is claiming a causality here that has not been proven. And even if it's true that his Tibetan monks are happier than other people, it still isn't necessarily due to the prevailing practices there. Maybe they entered the monastery already more content than other people; maybe Tibetan orders attract particularly well-adjusted people. That's why my colleague Tania Singer now wants to study specifically the effect of those methods the Dalai Lama recommends. She would like to instruct laypeople in a Buddhist meditation that is supposed to strengthen compassion and then measure how their brains, emotions, and behavior change with time. Maybe that will get us somewhere.

There have been wonderful experiments looking at the relationship between happiness and altruism in the other direction. They work something like this: I arrange for you to find some money. Afterward, your mood will be somewhat elevated. And it actually turns out that people are regularly more generous and helpful after an event like that.

When good things happen to people, they want to pass on the positive experience. As is so often the case, however, we lack a proper understanding of what exactly is happening in us in that situation. In general, I consider this to be one of the biggest unanswered questions of the social sciences: How does a society influence the individual? We know woefully little about that.

You've just described the problem in a very general way.

Because we run into it everywhere. Take our experiments with children. Of course, you immediately speculate that the degree of their willingness to share has something to do with upbringing on the one hand and with genetics on the other.

Then you can figure out with well-controlled investigations of twins, for example, how much genes contribute and how much the environment contributes.

Yes, but you still don't learn that much from that. Even if you managed to identify particular educational practices through which children turn out to be less selfish after a while, you still don't know how or why those practices work. What we find are almost always only particular correlations, far too rarely the true causes.

If we knew the causes, we might be able to teach children to be more altruistic.

To understand the mechanisms of altruism would be a necessary first step—but unfortunately no guarantee of success. Just think of how much effort parents put into instilling self-control in their kids. Some succeed; other people remain impulsive for their whole lives. And then I'm not even sure whether we ought to implant a deep sense of fairness in everyone at all.

Why not?

Because it's a secret of the success of the market economy that it separates certain processes from the sense of fairness. That's why we value property so highly. Imagine that you and I agree on an exchange granting you 20 percent and me 80 percent. That goes against your sense of fairness, of course. Yet new values might be produced through the exchange. . . .

For example, if I give you a tool you can use more than I can.

Exactly. New wealth is generated. But if you insist too strongly on your sense of fairness, you prevent the cake as a whole from getting bigger. And there's a second argument: We've investigated tribes in New Guinea whose members immediately have to share everything they get. They lack any incentive for effort. The principle of equality can hinder economic progress; in that regard, conservative economists are right. The question is only whether, contrary to their views, too much inequality eventually proves to be a drag on a society as well.

You've studied the sharing customs in more than a dozen different tribal societies around the world. I wonder whether a culture of hunter-gatherers in the jungle of New Guinea can actually be compared with that of Mongolian herders.

That very problem has confronted anthropologists up to now: They didn't have any measuring instruments that applied to everyone. But we had people play the same game everywhere. Even the wording of the instructions was exactly the same for all the cultures we studied. The behavioral experiment gave us insight into how much of a gift the players were willing to relinquish to others and what they considered a fair apportionment. The differences were vast. An extreme case was that of the Machiguenga in the Peruvian rainforest, who would part with virtually nothing at all. Those people in fact behave the way traditional economics actually expects all of us to behave. But the economic system of the Machiguenga is very primitive.

The noble savage is a myth.

Oh, yes. On the contrary, as we discovered, people share all the more willingly the more they're accustomed to trade and exchange. They know that it's to everyone's advantage for them to give a little sometimes, get a little at other times. And the market has taught them to compare the values of different goods.

That means: Even if something like a sense of fairness is innate to us, we still have to learn how to put it into practice.

Yes. And you can create environments that encourage people's altruistic tendencies—or stifle them. A colleague of mine compared two bicycle messenger services. In one of them, the messengers were paid per hour, while in the other, they were paid per delivery. In an experiment similar to the trust game, the messengers with the hourly wages proved to be far more altruistic than their colleagues who worked at a piece rate. Apparently, the latter had simply gotten used to the idea of every man for himself.

Haven't we all gotten used to that? The idea that we're all motivated by supposedly performance-based pay to do our best—hasn't that been the mantra in recent times?

It was certainly a mistake to give money such priority. Pay is an important incentive, but not the only one. Personally, I believe that the desire for recognition motivates us much more. For most people, there's nothing worse than being the loser whom no one takes seriously. Avoiding this fate is far more important to us than a somewhat higher salary—at least once a certain income level has been reached. A top banker will work seventy hours a week whether he's paid half a million a year or ten million.

We've been talking about money this whole time. But there are many other forms of fairness—for example, equality of opportunity or equality before the law. These barely figure in your research.

Unfortunately, we lack effective means of measuring those dimensions of fairness. Financial assets, however, can be compared very easily.

In that respect, your experiments are really true to life: We're constantly comparing how much we ourselves have with how much others have. In a society that's so obsessed with account balances, we shouldn't be surprised to find envy and resentment.

Yes, but it should be added that whether inequality offends our sense of fairness is extremely dependent on how it came about.

So, then, we could never define "fairness" abstractly—always only in terms of specific situations.

That's right. As soon as circumstances become only a little bit more complicated, the question of what is fair can be debated endlessly. For example, people are far more willing to share a gift with other people than something they acquired through hard work. The most unbearable feeling is that someone else got ahead by cheating us. To punish a presumed cheater people will even incur disadvantages to themselves. Here's a thought experiment: Everyone can put money in a pot to stick it to those top managers who led their companies into near ruin while enriching themselves. For every euro you invest, ten euros are taken from those managers and then burned.

Probably a sizeable fund would be collected. But that would be more an expression of revenge than of fairness.

Revenge is nothing but the dark side of the sense of fairness. It's a defense against the freeloaders in the community. As we were able to show in experiments, cooperation in a group usually breaks down very quickly when the group is afflicted with egoists. Only when the members of goodwill can punish the freeloaders does the cooperation become stable.

How did you come up with those experiments?

In an experiment, we once had a very large number of workers compete over a very small number of jobs, so that the workers had to accept very low wages and ultimately got angry—but instead of directing their ire at the employers who were taking advantage of the situation, they directed it at the other workers who were willing to work for a pittance and so forced their fellow workers to do the same. So each participant accused all the others of being scabs of a sort. We wondered how much the people would give to punish others for their uncooperative behavior.

It's interesting that all the workers play the game even though they find it unfair. Apparently, Bertolt Brecht was right: "First comes eating, then comes morality."

No one acts altruistically without regard to the costs. Remarkably, even people with strong fairness preferences will behave as if they didn't have any. They only have to get into an underbidding competition. Then all of them behave like perfect egoists.

The only way out would be if someone established trust. The players have to believe firmly that no one will stab anyone else in the back.

Of course, they could form a union. Unfortunately, collectives also often resort to nasty strategies to force cooperation. In an extreme case, you can set fire to the houses of the rich, as one of my colleagues in Russia and the Ukraine observed. There the farms of many peasants who had acquired some land and worked hard for a basic level of wealth went up in flames—not a rare phenomenon at all. Thinking in terms of fairness always has a destructive and a progressive side.

Maybe the human hunger for fairness is insatiable, and we're willing to sacrifice almost everything for it. Max Frisch once noted that all revolutionaries in history promised people justice, not happiness.

Happiness is a private good, justice a public good. Because you as an individual can do something for your well-being, the subject is ill-suited for revolutions. For justice, on the other hand, you have to fight together with others.

When does injustice cause a society to collapse?

When it fails because of its own ideology. The problem isn't inequality in itself, but whether you can justify it. In our culture, it's legitimized by performance. For decades people have heard that whoever works particularly hard or has special skills, and so benefits society, will be rewarded. But now it turns out that certain top managers hardly drew their vast income because of their excellent performance. Something like that can threaten the self-image of a whole society. We're not only in an economic crisis, but also in a moral one.

The Strongest Feeling of All

Neuropharmacologist
Walter Zieglgänsberger
on pain

◆

WALTER ZIEGLGÄNSBERGER IS THE MOST FAMOUS pain researcher in Germany. Born in 1940 in Landshut, Bavaria, he studied medicine in Munich, where, since 1984, he has led a research group at the Max Planck Institute of Psychiatry. We met at a Berlin exhibit that attempted to bring together the perspectives of art and science on pain. I viewed the exhibit with mixed feelings, because I found the representation of the science too superficial. Zieglgänsberger, meanwhile, who had traveled to Berlin from his Munich lab specifically for our conversation and museum visit, judged it even more harshly. For him, the many pictures of mutilated bodies and old surgical instruments had a lot to do with disgust and horror, but not much to do with pain. "The truly bad pain is the agony that begins very gradually—the slight ache in the back that gets more and more intense and over the years comes to define a whole life."

At one point, however, the scientist cringed. On the floor in front of us lay a rag doll in the form of a life-size horse carcass, covered in real horse skin. Zieglgänsberger, an avid rider, had recently had to put his horse to sleep: "That was very painful."

◆

Professor Zieglgänsberger, how long have you been studying pain?

For more than thirty years. As a very young scientist, I performed experiments on nerve cells in the spinal cord. In the process, we discovered that with some glutamate you could induce those cells to amplify pain signals. At the time, no one was prepared to believe that. Everyone thought that pain was nothing more than a message from an injury, and the worse the injury, the more intense the pain. What I was saying didn't necessarily advance my career. Well-known scientists chided us: What are these young researchers doing injecting soup seasoning into the brain?

You were contradicting a dogma. More than 350 years ago, Descartes, the French philosopher, compared pain with the ringing of a church bell. When someone pulls the rope down below, the bell tolls in the tower. Most people probably still see it that way. In one respect, at least, I like this image: Descartes understands pain as a process in our body—and not, say, as punishment for our sins.

Yes, that can be regarded as progress. But Descartes was still wrong: Pain is not simply a reflex. It depends a great deal, for example, on the context in which you experience the pain. To gauge the pain threshold of test subjects, we place a small piece of metal on their skin and heat it. When I myself did this, the subjects would complain at a later point than when my assistants performed the test. The mere presence of a professor has a pain-relieving effect.

Maybe they were simply less comfortable whining in your presence.

No. They actually felt less pain, as we were able to measure by their brain activity. You can't account for any of that with Descartes.

I was impressed by experiments conducted by your Heidelberg colleague Herta Flor on people with chronic back pain. Many of them had relationships with empathetic partners. Others had life companions who didn't want to hear about their suffering. But the latter patients were actually better off: It turned out that attention and words of comfort increase pain.

The system is simply enormously adaptable—here it reacts to a well-meaning partner. We had to radically rethink our views. Just a few years earlier, we had thought we were sure that everything is hardwired in our heads forever. Now we know that even an adult brain is constantly transforming. For us neuroscientists, this is an incredibly exciting time—as if we were just now grasping that the earth isn't flat. By the way, we now also suspect that an important role is played by glutamate, which is produced by the body itself: Under its influence, neurons can adapt more easily.

Are there also innate differences in how intensely we feel pain?

Yes. Sensitivity to pain is hereditary. And there are even people who are, as a result of their genetic endowment, completely insensitive to pain. Gender also plays a role. Most women are more sensitive to pain than men. That has to do with the female nervous system. Red-haired women, however, have an advantage. They have a genetic variant that makes them particularly responsive to certain opioids.

Those are the pain-relieving and intoxicating agents in opium. The body itself produces them as well. We all have a private drug lab in our brain, so to speak.

Well, these opioids primarily serve the purpose of inhibiting pain in a natural way. Without this system, a wounded animal being pursued by a predator would be doomed. Injured soccer players also have these opioids to thank for the fact that they can keep playing. They're the strongest painkillers in existence.

Amazingly, the power of imagination is enough to release opioids. If pain patients are given a placebo that they firmly believe is an active drug, their brains release opioids—and they feel better.

That's right. You can prove that placebo effect by their brain activity as well. Whether the painkiller comes from outside or inside ultimately doesn't matter. With techniques like meditation and yoga, you can induce the brain to use its own substances. Fakirs sit on beds of nails without taking pills beforehand.

What actually causes the perception of pain?

Almost everywhere on your body, there are sensors known as nociceptors, which respond to heat, intense pressure, or chemical stimuli. When you're injured somewhere, the nociceptors send a signal to the spinal cord, where something very important happens: The pain signal takes precedence over all other messages from your body. In addition, the information is divided. One channel leads into areas of the cerebrum that localize the injury. Other impulses enter the deeper brain regions and trigger the unpleasant sensations.

There are people who find pain pleasurable. Two million—by some estimates up to six million—Germans take pleasure more or less regularly in masochistic games.

It's not pain that those people feel. They just find their practices exciting. You can see that in something as ordinary as sexual intercourse. Recently, friends of mine returning home from an evening out found their children crying on the sofa. The kids had watched a soft porn video their parents had in the cabinet. "They were naked, and the man was hurting the woman so badly," they said. The parents didn't know what to do other than watch the movie with their kids again. The scene that had frightened the children so much showed a completely standard missionary position. But the woman's face was actually contorted as if she were suffering terrible pain. The kids interpreted that correctly. Only we don't notice it.

Because the excitement is more intense than the pain.

Above all because we can control the situation. Even in the most extreme masochistic games, you need only say a code word, and the scene is ended. The fear that first makes pain so intolerable is absent.

Pain is powerlessness.

That's what first makes it distressful. This is due to the fact that the pain information is divided. If you observed the brain during sex, you would see that the regions for localizing pain certainly respond to an excessive pressure, maybe even to a slight tissue injury. But because the situation is clearly harmless, the centers for unpleasant emotions send hardly any signals. Otherwise, we wouldn't keep doing it, of course.

You once asserted that dangerous pain isn't intense, but faint.

The body is prepared for an acute injury. Opioids are released; in the worst-case scenario, you pass out. But a nagging, recurring back pain, say, always slight, slight . . . it can drive people to despair. All day long you're focusing on whether it's coming back. Gradually you get anxious. And then something fatal happens: You become more and more sensitive to the pain, because your nervous system is beginning to change. When the neurons are stimulated again and again, they amplify the incoming signals more and more. That's how the brain learns—through repetition. In this case, it unfortunately learns pain. In that way, the initially slight sensation becomes stronger and stronger. In the end, the pain defines your whole life.

And the back? What's wrong with the back?

Most of the time, nothing at all. The pain itself is the disease. Maybe you've had muscle tension at some point. That's how innocently a lifetime of pain can begin. That's due not least of all to doctors who cause anxiety. Because they don't know for sure, they make diagnoses like: "Well, your back, it's not the strongest. Even if you don't have any serious problems now, you probably will in twenty years." Out of fear, patients begin to pay attention to every little ache in their back. In that way, they program themselves for pain. And because they believe they have to go easy on their bodies, their muscles weaken and tense up all the more at the next opportunity.

There's even a name for that process: iatrogenic, which refers to an adverse condition caused by a doctor.

That actually occasionally borders on bodily harm.

Modern medicine promotes developments like that. Nowadays, thanks to ultrasound, CT scanning, and genetic testing, doctors see abnormalities that previously would have been completely hidden from them. Are they supposed to keep silent about them?

We doctors should at least watch what we say—especially when we're actually looking at healthy people. Language is the sharpest of all swords. That's why I'm always extremely careful when I'm explaining things to a patient.

In recent years, patients have increasingly been diagnosing themselves with information from the Internet. Of course, I get uneasy when I read that my knee pain could be associated with a bone infarction. But do you want to turn medicine back into a secret science? It seems to me to be more a matter of learning to deal with this knowledge.

For that very reason—because patients can inform themselves so easily—we doctors shouldn't engage in groundless speculation. Our job is to give them hope. By the way, it's advisable to prescribe painkillers—even strong ones—as early as possible so that the learning process can't get underway in the first place.

What do you do once the pain has become chronic?

What people have learned, they can unlearn. First we have to turn off the pain. For that, you usually need medicine. But that's only a crutch. In the past, we made huge mistakes in that regard. As soon as we had the patients pain-free, we often left them alone. But if we don't undo the programming of their nervous system for pain, the torments return. Unfortunately, the brain doesn't have a delete key.

We get over a bad experience only when we overwrite it in our memory with a better one.

The patients withdraw more and more from life with their pain. If I were, say, an opera lover and became afflicted with chronic pain, the next time I saw *Tristan and Isolde*, the very thing I used to love would now be a torment to me. By the time the two of them died for love at the end, I would have seen myself die three times from my pain. From then on, opera would be invested with negative associations. At that point, all that could help would be to rediscover how beautiful Wagner's music can be. To the extent that patients regain their joie de vivre, they lose their anxiety and stop focusing obsessively on their suffering. Going too easy on your body doesn't get you anywhere. You're better off rebuilding your capacity for joy.

To cope with chronic physical pain, you need a sort of psychotherapy.

Yes. The key is to overcome the anxiety about it. You can't do that with medicine alone.

The flip side is that a lot of what we have learned about physical pain applies equally to mental anguish. The brain doesn't make any distinction: When we suffer from lovesickness, grieve over a rejection or even just a financial loss, the same circuits are activated in the head as when we feel physical pain. "You've really hurt me," we say in those situations. Sometimes our everyday parlance is impressively precise.

Much more precise, actually, than our traditional philosophy, which still sees a fundamental difference between physical and mental phenomena. But when I clap my hands, does my left hand or my right hand make the sound? It's exactly the same with pain of the body and the mind. Psychological suffering can destroy you too.

In our society, a different view has prevailed until now. If you hit a child, that's abuse. Threats to withdraw your love, however, you can make as viciously as you please. And in the professional lives of adults, it's almost a prerequisite that you master the techniques of workplace bullying.

Here, too, powerlessness is the actual problem. Children suffer much more from emotional pain than from a slap to the cheek. They feel as if they're at the mercy of their parents—just as victims of workplace bullying feel as if they're at the mercy of their colleagues. That's what first makes the experience so unbearable.

People's capacity for empathy can prevent them from such acts of cruelty to others. When we witness the pain of others, our brain reacts to it just as if we ourselves had been hurt.

In my lectures, I sometimes show a short video. It's a clip of the tennis pro Michael Stich lunging and tearing a ligament. His ankle twists until the sole of his foot is facing upward. . . .

Owww!

See, you're suffering vicariously too, even though I've only told you about it. Yet you're presumably not very close to Michael Stich.

Our ability to experience so directly even a stranger's pain as our own is completely at odds with our customary view of humanity. We usually imagine human beings in their natural state as creatures without scruples who think only of their own well-being. The fact that we don't beat each other's heads in is regarded solely as an achievement of culture and education. But as new discoveries in neuroscience show, we most likely have an innate capacity for compassion.

Human beings are anything but altruists. That's exactly why we need the ability to feel the pain of others, not to mention their overall mood. Without empathy our social life probably wouldn't work.

So how do you as a doctor tolerate witnessing the agonizing pain of your patients?

It's a balancing act. You have to be able to put yourself in your patients' shoes to some extent, or else they won't accept you as a doctor in the first place. On the other hand, you need a professional distance, or else you'll be devastated. What helps is that the patients love you when you relieve their pain. Galen, the famous doctor of antiquity, already knew that: "It is divine to allay pain."

How did you come to study pain, of all things?

I wanted to understand the signal processing in the nervous system. Pain was simply a good example.

It was that abstract?

That's how I saw it as a young scientist. Only with time did I learn how much these things have to do with completely everyday life.

Your other major research topic is addiction.

Pain and addiction are very closely related. Both are learning processes gone wrong. With chronic pain, the brain learns to trigger stronger and stronger sensations that no longer benefit the organism at all. The addicted brain has programmed itself to need ever-greater amounts of drugs in order to function just semi-decently. Both processes are based on comparable modifications in neural networks. And both are real diseases—even though a lot of people still refuse to believe it.

Neuroscience has changed our understanding of humanity so quickly that many people find it hard to go along with it. They resist seeing their most personal moods and emotions reduced to a complex set of physical changes in their heads.

It's not only philosophy but even more so our everyday culture that still separates body and mind. But we now know that the majority of cases of depression, for example, are not a logical response to a life situation but rather metabolic disorders. We're really developing something like a molecular psychology. I approach the difficulties many people have with that more from a practical angle: If I want to persuade my patients to take medication, I have to consider their sensitivities. I tell them that I understand their antipathy and might even react the exact same way under similar circumstances. I recommend an antidepressant as a chemical crutch to help them get out of the vale of tears. They can throw them away as soon as they've stabilized.

Particularly when I'm suffering from intense pain, I find it hard to understand my body and my experience as a unity. At those times, I feel more like my own body has turned against me—whatever "me" is. It's my enemy; I want to break free from it.

But that, too, is taking place in your brain. There your body is reproduced as if on a map. And if your knee hurts repeatedly, more and more nerve cells receive that information over time. So the map of the body changes: The part that's in pain gets bigger and bigger on it.

It's a long way from the changes in individual brain cells to sensations and human behavior. How do you bridge that gap?

If you're asking me as a scientist: We're currently working on methods of investigating the interactions of larger and larger neural networks. In that way, we hope to better understand how processes between cells are translated step-by-step into human feelings and impulses. If you're asking me as a doctor: Up until a few years ago, alongside my lab work, I was often on call as an emergency doctor at night and on weekends. I needed the feeling of doing something immediately worthwhile. In the process, I came into contact with pain and addiction in extreme situations.

How did the director of your institute react to your nighttime activities?

We had a deal; it stays between us. Even when I admitted patients to our own hospital, my colleagues probably always thought it was my brother, who has his own practice. I still benefit from that experience. The doctors see me as one of them, not an out-of-touch researcher—I speak their language.

At least two thousand people in Germany kill themselves each year because they can't bear their pain. Do you think that treatment could help all of them?

There are probably more victims than that: You have to add all those who want to be delivered from a terminal illness. In reality, they only want to get rid of their pain. We know that good pain treatment dramatically reduces the desire to die. With new treatments, we can address cases that were previously considered hopeless. That's the goal with which I've approached my research: We have to try everything to lessen suffering and maintain joy in life.

Everything? The idea that pain has a meaning and we should endure it is deep-seated in us.

Constant pain has no meaning. We should forget as quickly as we can those mystical ideas that suffering ennobles people. That's an antiquated perspective in Christianity. Even the pope has declared his support for pain treatment. No one should have to bear pain.

Can there be a world without pain?

Pain will always be a signal—even if an extreme one—that we're alive.

The Female Side of Evolution

Anthropologist

Sarah Hrdy

on motherhood

◆

NOT MANY PEOPLE HAVE THOUGHT as much about what it means to be a mother as Sarah Hrdy has. In this emotionally charged domain, the American anthropologist is considered an authority. But she herself describes the focus of her career with a strange distance, as if she wanted to steer clear of any hint of sentimentality: She sees herself as a "mammal with the emotional legacy that makes me capable of caring for others, breeding with the ovaries of an ape, possessing the mind of a human being." But Hrdy is anything but a cold person, and she must have been plagued often enough in her own life by the question she poses: "What does it mean to be all these things embodied in one ambitious woman?"

Born in 1946 in Dallas, Texas, she studied anthropology at Harvard and made a name for herself with research on the social life of the langur monkeys on the holy Mount Abu in North India. In 1984 the University of California, Davis, appointed her professor of anthropology. But she left her academic career when she was only fifty, "eager to continue research and writing while seeking a more harmonious balance between work and family." Currently, she lives on a farm in northern California, where she grows walnuts and writes books.

We met at Cambridge, England, on the occasion of the Darwin anniversary festival. Hrdy spoke softly and hoarsely, now and then struggling for control of her vocal cords. Several lectures on the female side of evolution had nearly cost her her voice.

◆

Professor Hrdy, how do you remember your mother?

She was a very beautiful, very smart, and very ambitious woman.

You once described her as "by some standards, an appalling mother." What did you mean by that?

Well, her position in society was more important to her than her kids. For us, there was a constantly changing succession of nannies, plus she subscribed to the view then common among educated mothers that babies were born blank slates, needing to be shaped. Picking up a crying baby would just spoil her, conditioning the child to cry more. Unquestionably though, I loved her and later in life felt very close to my mother, having learned to understand the constraints she herself confronted and how she herself had been reared—in a long line of mothers gradually losing the art of nurture.

Your books give the reader the impression that no human relationship is anywhere near as fraught with tension as the relationship between a mother and her children.

Under some circumstances, a mother can devote herself completely to her children. But often the mother herself lacks support, or she has to divide her love among several children, or she has other things she needs to do. Human mothers have always confronted such trade-offs. Maternal ambivalence is as natural as maternal love. Yet it's been hammered into us that unconditional and self-sacrificing maternal love is "normal," ambivalence considered pathological. It's assumed that mothers should readily turn their lives over to their little "gene vehicles". . . .

As some spiders do.

Think of *Diaea ergandros*. As soon as the young of this Australian species have hatched, the mother undergoes a strange paralysis. She then secretes a substance that liquefies her own body. The mother turns herself into an edible slime—the first nourishment for her offspring, and it's a good thing, too, since being less hungry, the spiderlings are less likely to cannibalize each other.

You yourself have two adult daughters and a son. Are you familiar with that feeling of paralysis—and the fear of being annihilated by your own children?

Oh, yes. But it's not only as a mother that I've felt as if my family might eat me alive; I sometimes felt that way as a daughter as well. The Texas I grew up in was still extremely patriarchal, not to mention racist. Of course, as a young girl I didn't have the faintest idea of what this fixation with controlling girls was about. It was the same for my mother: She wanted to be a lawyer, but my grandmother insisted that she first make her debut in Dallas society. There she met my father—a great catch, heir to an oil fortune. And that was that.

You don't lead exactly the life people expect from a wealthy heiress of several oil companies. How did you break away from that background?

Well, as the third daughter in a family desperate for a son, I was the heiress to spare. Because I loved horses, I was allowed to go off to a boarding school known for its riding program. Fortunately for me, that school also took women's education very seriously. From there,

I went to Wellesley College, where my mother and grandmother had gone. I embarked on a novel about contemporary Mexicans of Maya descent, and it occurred to me that it would be a good idea to learn more about Maya culture. So I transferred to Radcliffe, then the women's part of Harvard, to study under the great Mayanist Evon Vogt. The novel never got finished, but I ended up as an anthropologist. It's hard to believe how naive I was, but after summers working in Guatemala and Honduras as an undergraduate volunteer on medical projects, I gradually came to understand just how oppressive the political situation was. Continuing as an anthropologist in that part of the world would turn me into a revolutionary—something I was temperamentally unsuited for.

Instead you went to India in 1971 to study langur monkeys. What drew you there?

I vaguely remembered from an undergraduate course that there was this species of monkey in India called langurs, and that, supposedly due to crowding, the males would occasionally kill babies of their own species. Naively, again, I imagined that langurs would provide a scientific case study for how overpopulation can produce pathological behavior. By the end of my first field season, I realized my starting hypothesis was wrong. Female langurs live in groups with overlapping generations of female kin accompanied by a male who enters from outside. Every so often, a new male from one of the roving all-male bands manages to oust the resident male and take his place. Babies were only being attacked when new males entered the breeding system from outside.

So the intruder attacks the babies?

Exactly. By killing the offspring of his predecessor, the usurper reduced the time before the no-longer-lactating female ovulated again. Far from pathological behavior, the male was increasing his own chances to breed.

The babies' mothers . . .

. . . try to evade him. But the male has the advantage of being bigger, stronger, and incredibly persistent, stalking the mother-infant pair day after day. Occasionally, a young mother would simply give up trying to protect her baby. After its death she will sexually solicit the murderer. To successfully reproduce, females sometimes have to be opportunistic.

Were you able to watch dispassionately?

No, I was distressed. Watching an attack, tears would roll down my cheeks.

Scientists are not supposed to intervene in the process they are studying.

Right. Plus even if I had tried, I could not have stopped what was happening. And yes, I was also fascinated. After all, this was the bizarre phenomenon that I had come to India to try to understand—why were males doing this? Why, instead of sexually boycotting the male who killed the infant, were mothers going along with it?

The killing of babies was conclusive evidence that what happens in nature can be even harsher than we thought. At the time, many people understood Darwin in terms of the idea that in social animals like primates, all the animals behaved so as to promote the survival of the group or the species. According to your findings, however, even among members of the same species, each individual is striving to maximize its own reproductive chances—even at the cost of its own offspring. But why would any mother then mate with the same male who had killed her infant?

Because a female who boycotted the infanticidal male would be at a disadvantage in relation to other females in her group who bred faster. Plus, to the extent that the infanticidal trait was heritable and advantageous, her sons would inherit it.

Sociobiology, which emerged at that time and advocated such theories, was not quite welcomed with open arms.

That's putting it mildly. Back in the 1970s, opponents of sociobiology characterized it as sexist, racist, and worse—practically Nazi! We described the world in a way that was out of step with the times. Worse for me—because I was a sociobiologist among anthropologists who distrusted evolutionary perspectives, and was also trying to introduce a female-oriented perspective among evolutionists who regarded feminism as hopelessly suspect. I was decried as a double ideologue. However, in my opinion it's simply better science to take into account Darwinian selection pressures on mothers and infants, and biased science to leave out half the species.

No one can accuse you of that, at least. You've studied female reproduction in all its stages—beginning with the sexuality of the langur females.

At that time, the conventional wisdom was that females mate only when they're ovulating and only with one male. For many animals, that's true. But I noticed that langur females solicit males for sex even when they're not fertile and would sometimes leave the troop to solicit outside males as well. Why? No one before had bothered to wonder about these multiple matings, but it has become an important subject of study.

Were you intrigued by this question because you thought the answer would apply to us humans as well?

As interest in the topic grew I was actually invited to speak on non-conceptive sex in nature at a conference organized by the Italian government and the US National Science Foundation about the "meaning of sexual intercourse," and yes, they clearly had humans in mind. The Vatican also sent representatives. What a tragedy for Catholics and the world that Saint Augustine, on whom the Church bases its policies to this day, was such a poor animal behaviorist! Augustine believed that he could extrapolate from domestic animals to all animals. For many of them, sex actually is confined to the period right around ovulation when conception can occur, but this is hardly universal in nature, and certainly not among many primates.

So what purpose does it serve for the langur females?

Maybe to manipulate information available to males about paternity. Males apparently remember whom they mated with and avoid harming offspring that might just possibly be their own. But how to make females solicit multiple matings? This led me to speculate that the female orgasm—which, by the way, definitely occurs in some other primates as well—originally evolved long, long ago, in a prehuman context, as part of a reward system for multiple copulations. Women have inherited a physiological potential whose workings evolved in a prehuman social and reproductive context very different from that found among most humans today.

You have to explain that.

To mate with multiple partners is a dangerous business for langurs, too. What can motivate a female creature to do that? We're talking about a pleasurable psychophysical reaction in response to repeated stimulation, which, on top of that, is erratic. Actually the most powerful type of conditioning is when the reward is uncertain. Thus nonhuman primate females were conditioned to solicit successive partners. I doubt that the pleasurable psychophysical reaction that we call orgasm evolved originally in order to promote human pair bonds. If it had, frankly, we would expect the response to be more reliable, to function better than it does. Rather, the female orgasm is far older, a vestige, secondarily conscripted to this new function.

You mean a woman's climax is as superfluous as her appendix?

Perhaps, or a still-detectable vestige of a former adaptation no longer so strongly selected for—like the grasping reflex of our newborns for the nonexistent fur of the mother. If I'm right, our descendants in spaceships eons after us may wonder why we made such a fuss about this subject.

It's risky to draw conclusions about humans on the basis of langurs. Can you prove your theory?

No. It's a hypothesis, and a highly speculative one at that.

In any case, you women pay dearly for your desire. For the sake of reproductive success, a woman can at best use her libido to secure for herself the support of as many men as possible—by keeping them

well-disposed toward her with sex and obscuring her fertility. But the better she manages to do that, the more uncertain each man is of his paternity, the less he does for his offspring. The way out for the woman is probably to find an optimal balance between boundless promiscuity and strict faithfulness.

Well, that balance can be different for each species and, among humans, for each culture, and for women in different economic circumstances. Some South American tribes with unpredictable resources, where mothers have no guarantees that a man will be able to provide for his children, or even still be alive or around, have come up with a fascinating cultural solution to an ancient biological dilemma. They subscribe to the belief that during pregnancy the fetus is built up from semen contributed by each man a woman has sex with during that period. These contributors are all considered "fathers" and expected to help provision the mother and her subsequently born child. Research among the Bari people of Venezuela by the anthropologist Steve Beckerman has shown that children with several putative fathers are more likely to survive. *Too many* fathers could be detrimental, presumably because men felt less responsible. The babies' best chances of survival actually turned out to be two supposed fathers.

Men in our monogamous culture would be less than thrilled. But that calculation can hardly be applied to a highly developed society, in which almost every baby survives.

Right. Plus, in our society, with a long, long tradition of patrilineal inheritance of property, we have reified and glorified this concept of paternity certainty. But what's interesting is how much variation there is in how willing human males are to invest in offspring. A few years ago, *Time* magazine asked me for a brief article for Father's Day. I wrote that there's a huge potential for male nurture in the human species, albeit a potential not always tapped. In passing I cited some statistics from the Children's Defense Fund showing that divorced fathers were more willing to make car payments than pay child support. I got some pretty angry e-mails, and a website was set up for people to directly complain about me to the magazine's publisher.

The men's behavior isn't right or justifiable—but I can understand it.

Of course: If you don't have a car, you won't get another woman.

Aren't you taking sociobiology a bit too far now? It's possible that some men are quite simply angry with their ex-wives and take out their anger on their children.

Yes, of course. But such anger has a deep evolutionary history as well: The competition between men for women; the attempt to control women and constrain their choices, along with women's efforts to evade such control—we owe this to Darwinian sexual selection. The great biologist E. O. Wilson asserted that sexual selection is the most antisocial force in nature. Ultimately, the male strategies and female strategies I was studying among the langurs were the products of sexual selection.

While doing your field research on langurs in North India, you yourself had a baby with you, your little daughter, while your husband was working in Boston. How did that work?

I was going back and forth to Rajasthan. I hired a babysitter to accompany us, but the logistics were difficult, and there were diarrhea and diaper rash problems. I had specifically instructed the sitter to never give our baby, Katrinka, food near the monkeys, but of course one day she handed her a cookie, and the langurs were all over them, frightening Katrinka. Even worse were the political problems we encountered working in India at a time when the Nixon government was "tilted" toward Pakistan. Thereafter I decided to work closer to home in ways less disruptive to my husband and children.

Why did you want to have kids?

Not sure! But I am glad we did.

Back at Harvard, Robert Trivers, one of the fathers of sociobiology, criticized you in an interview, saying, "Sarah ought to devote more

time and study and thought to raising a healthy daughter." Did his words hurt you?

Worse, I feared that Bob Trivers might be right! Almost every mother agonizes over the possibility that they are not doing enough.

Trivers was actually making a typical sociobiological argument: By nature, each organism exists only for reproduction. Status acquisition is secondary.

But in fact, from an evolutionary perspective, it's impossible to separate a woman's striving for status and her reproductive success as a mother. Mother Nature, my personal metaphor for Darwinian natural selection, motivated females to compete for resources and to solicit sex, not to have an instinctive desire to have children. Natural selection instilled sufficient ambition to compete for local resources that the female needs to mature and stockpile reproductive fat. Since sex is a matter of course, any female fat enough to ovulate will conceive and bear children. Female primates were selected to strive for local clout so as to improve their chances of bearing young, and after offspring were born, to be able to protect and provide for them. It's no surprise that competing for status can be a higher immediate priority than having children.

Or prevent it entirely.

Yes. With the availability of safe and reliable birth control, women can vote with their ovaries, opting for improved status or subsistence while delaying childbirth or avoiding it altogether, or delaying until they find themselves in a more supportive situation.

Which is enormous progress: In the past, women who were at the end of their rope abandoned their children—as in the tale of Hansel and Gretel.

Such behavior was much more frequent than most people realize. In Tuscany, for example, between 1500 and 1700, at least 12 percent of all children were left at orphanages. The largest foundling home in Florence by itself took in up to five thousand children a year, even though mothers would have been aware of dismal prospects for a foundling: In Sicily, for example, as recently as the nineteenth century, only 20 percent of abandoned babies survived. Residents of the northern Italian city of Brescia proposed that a motto be carved over the gate of the foundling home: "Here children are killed at public expense."

So our species doesn't have much right to point fingers at baby-killing langurs.

There's a big difference between what happens in nature and what is "morally" acceptable. But beyond that, the patterning of infanticide in humans is rather different than in langurs: Human mothers may abandon their babies when they feel incapable of supporting them. Some even commit violence against their child. But virtually no nonhuman primate mother in nature has ever been known to abandon or inflict damage on her own infant. Other primate mothers will care for even the most severely disabled baby, and carry dead babies for days rather than give them up. A langur mother submits to violence when a new ruler takes over her group, but she has no choice: Sooner or later the invader will succeed in killing her infant, and if she didn't breed with him, she would lessen her chances of conceiving again in a timely fashion.

Are humans more inhumane to their children than monkeys? Listening to you, someone could get the impression that you see human motherly love as nothing but a sentimental myth.

The idea of the mother's unconditional commitment to her child is a myth, but the powerful commitment and protectiveness human mothers feel toward children is very real. The point is, maternal devotion in other primate mothers is more single-minded and does not depend so much on the mother's perception of social support—in that sense it's more reliable. Human maternal love is more variable and contingent, or dependent, on circumstances, especially in the window right after birth, the first seventy-two or so hours. Before the mother begins to respond to cues from her baby and lactation gets going, the emotional cost of giving up her baby is far lower. Many cultures take that into account, not considering the baby fully human until the baby begins to suckle, is given a name, ascribed a "soul." Prior to that social acknowledgement, they would not regard infanticide as killing another human. The human mother's bonds to her infant are less an automatic instinct, like a switch one turns on or off, than a process and an emergent response to her infant over days and weeks. In time she becomes literally addicted to her baby's appearance and smell. As measurements of brain activity have shown, the mere sight of babies' faces stimulates the so-called reward system in human brains—not just in mothers but in fathers and even men and women who have never had a child.

So a button nose and little chubby cheeks have the same effect as sex or cocaine.

Well, not exactly the same. But yes, the same neural receptors are involved. That's one reason addiction interferes so terribly with maternal responses.

Have your kids made you happy?

At the moment, they're doing fine, so yes. But I am keenly aware that a mother is rarely happier than her least happy child.

I'm familiar with this experience myself—even though I'm a father.

The difference is less significant than most people think. A father's hormone levels change, too, when he holds a baby. As early as 1980, it was found that increased prolactin circulated in the blood of male caretakers among marmosets. Marmosets belong to a subfamily of New World monkeys in which mothers rely on fathers and other group members to help them care for and provision their young. Up to that point, prolactin had been considered a female hormone, which regulates milk production, among other things. But since then we've learned that prolactin also rises in fathers in intimate contact with new babies, not just in mothers.

Coincidentally, I'm currently in that very situation, except I haven't noticed any hormonal changes yet. I wouldn't know what those changes are supposed to trigger in my body either.

The mechanism is unknown. And the increase is less marked than in women. On the other hand, we know that men with an elevated prolactin level after the birth react more attentively to crying babies. On top of that, testosterone levels in men nurturing babies goes down. Perhaps we shouldn't be surprised that it's men in societies where they are segregated from mothers and young children who tend to be the most hypermasculine and warlike.

Maybe German society would become gentler if all fathers took the parental leave prescribed by our minister for family affairs.

I think that it would—even if I can't prove it. Perhaps your Ministry for Family Affairs should pass out earplugs to married mothers at every birth.

Why?

Because women by nature have a slightly lower threshold for responding to a baby's cries than men do. Even if both parents have the best intentions, she will get there before him to comfort the baby. Thus the baby becomes more attached to the mother than the father. In that way, a slight genetic difference between the sexes in the threshold of responding to infant cries develops into a great asymmetry. With the earplugs, women level the playing field; and the more the man takes care of the baby, the more sensitive his antennae will become. Of course, using earplugs assumes that the mother can count on the father to be there.

It can't be due only to hearing that men typically take care of their children less than women.

Of course not. The great mystery is why some men do so much and others so little. Subsistence patterns and residence patterns seem to play significant roles. Among hunter-gatherers, for example, a couple was often still living among the mother's kin when a young woman first gave birth. But residence patterns in foraging societies tend to be opportunistic and flexible. Later on, the pair may move with their children to live among his relatives, or to some other group altogether. If maternal grandmothers, aunts, and nieces are around, men tend to do much less direct infant care. But in situations when other help is not so available he does far more—in one study by the anthropologist Courtney Meehan among the Aka central African foragers thirty times more! Apparently, those fathers do fill the breach, but when they have to.

> You've speculated that the most natural form of child rearing for humans is cooperative breeding: *Homo sapiens* could evolve only because parents didn't look after their children on their own, but many others contributed care.

Humans produce the costliest, slowest-maturing youngsters on the planet. Under the uncertain conditions with wildly unpredictable rainfall and resources that prevailed in Pleistocene Africa 1.8 million years ago, neither a single mother nor a single couple could have consistently procured sufficient food for offspring so dependent for so long. Shared care must have been the precondition for our species to be able to afford the luxury of such long childhoods and big brains.

> With that argument, you turn the traditional theory of how modern humans evolved on its head: Usually it is supposed that our ancestors developed language and reason before they could cooperate in a group.

Right. Among other apes, mothers never allow others to hold their baby and the only provisioning of young is via mother's milk. Orangutans, for example, nurse far longer, but once weaned, youngsters provision themselves. Human children go on depending on provisioning from others for years after weaning. What emerges from my research is that a big brain requires care far more than caring for others requires a big brain. Marmosets have tiny brains and little capacity for mentalizing about what others are "thinking," yet they are remarkably effective at coordinating with mothers to care for and provision their young—even young not particularly related to them. The fact that humans do so, too, might explain why we and not the chimps now rule the world. We had to become the nicest apes before we had a chance to become the smartest.

That sounds completely different from the usual sociobiology with its "struggle of all against all." According to that view, cooperation outside of the immediate family is hard to achieve, because each individual wants to clear the way for his or her own genes. What changed your mind?

You're right. Years ago, the great economist Adam Smith and Darwin himself noted the peculiarly other-regarding and cooperative impulses of humans, but we modern Darwinians were so distracted by male-male competition and so forth that we glossed over it. But competition between members of one sex for access to the other simply can't account for everything. The shift in focus to just how much nicer than other apes humans can be—so different in many respects from chimpanzees—is causing quite a ruckus in the field. There's a tendency among some to view the focus on other-regarding impulses as a Pollyannaish donning of rose-tinted glasses. So it's lucky for me that my early reputation was based on my research on infanticide. Clearly I am perfectly willing to acknowledge the potential for violence lurking in primates—primates aren't necessarily or always helpful to others belonging to their same species or group. But humans also do have this extraordinary potential for nurture—even nurturing the young of others.

But it might be difficult to find support for your hypothesis. We don't have any evidence for how humans lived more than a million years ago.

No. But we do have some clues. I propose that our ancestors already had a long childhood before they developed their big brains. The conventional theory of human evolution claims the exact opposite: First big brains, then long childhood. Both possibilities are consistent with the currently available fossil evidence. I also assume that our forebears left the forests when the climate changed about 1.8 million years ago. At that time, it became harder to feed their offspring. Cooperative breeding reduced the risk that a child would die: Whoever had food at a given time could share it with others who had none. And it freed the mothers to forage. We know from research on birds that cooperative breeding is correlated with significantly longer periods of post-fledging dependence, essentially longer childhoods. Among species that rear their young, the young are provisioned longer, buffering them from starvation at a very vulnerable time. Cooperative breeding also might have made it possible for human women to give birth to children at shorter intervals. Lower child mortality, more time for learning, and more births amounted to an evolutionary advantage—humans became more fertile.

> Nowadays, molecular geneticists can actually assess the size of the world population in the past. As it turned out, during most of prehistory, our ancestors were anything but numerous—the total human population may have included as few as ten thousand breeding adults. According to those estimates, our forebears didn't breed successfully at all. Chimpanzees, for example, did much better. Doesn't that contradict your theory?

Not really. The reason the genome of the world's remaining hundred-thousand-or-so chimpanzees contains so much more variation than that of all six-to-seven-or-so billion people on earth today is that there are these population bottlenecks in the past. Experts debate over how many bottlenecks, when, and how severe, but there is no question that human populations in the past were small and widely spread across vast areas of sub-Saharan Africa. At one or more points in our deep past, humanity barely escaped complete extinction. But the cause doesn't necessarily have to be low fertility. There may have been population crashes due to recurrent famines brought about by fluctuating rainfall in Pleistocene Africa—a type of environmental challenge that in many other species is correlated with the evolution of cooperative breeding. The savanna-woodland habitats our ancestors inhabited were far more challenging environments in which to rear young than the more stable habitats forest apes relied on. Certainly adverse environments were a far more chronic threat to early hominins than intergroup warfare. That's why I consider shared care and provisioning of young a more plausible explanation for the *initial* emergence of our other-regarding, helpful impulses than the need to cooperate with other group members in wiping out other groups—something we only have evidence for relatively late in the evolution of our species.

You're alluding to theories according to which cooperativeness and altruism arose from the propensity for violence. Ultimately, the advocates of that view are attempting to solve the same mystery as you are: They want to explain how the initial pure competition of all against all for the best reproductive chances was at some point overcome. If two groups are feuding, the one in which members put their necks on the line for each other will have better prospects. Our species could have developed a tendency toward unselfish behavior from that evolutionary advantage as well.

But only theoretically—and not if humans were living at low densities in widely dispersed groups whose membership fluctuated as individuals moved between them, gravitating away from adversity and toward opportunities. Why would a few scattered clans with no property or fixed territories to defend risk waging war when they could just move? Why kill people and steal their women if by doing so men eliminate or alienate the very alloparents [individuals who act as parents for nonbiological offspring] whose help is needed for their offspring to survive? Ancestral humans had more pressing priorities—like keeping youngsters alive in the face of very high rates of child mortality.

And how, in your view, would shared child care have created a cooperative and intelligent species?

The social context in which immatures develop is the key. Finding themselves dependent on multiple attachment figures, these infants needed to monitor others and elicit care from them, far more so than would a chimpanzee infant who could count on the single-minded devotion of his or her mother. From a very young age, these little hominins had to be able to mentalize the likes and dislikes of others, appeal to and ingratiate themselves. Over the course of development, a very different ape "phenotype" emerges, subsequently subjected over generations to directional Darwinian selection favoring any little ape just a little better at understanding someone else's perspective, and so forth. That infant would be more likely to be cared for, better fed, and ultimately more likely to survive. In that way, competition for care developed among children, gradually leading to the evolution of greater mind-reading abilities. If you compare the cognitive capacities of a two-year-old today with that of a chimp, you won't notice very significant differences—except in one realm. Human toddlers are much better at taking the perspective of others, learning from them, and ingratiating themselves by sharing. And that was exactly what mattered.

If your theory is correct, our current, isolated nuclear families would be perverse. They demand too much of the parents and at the same time make it harder for children to develop their capacity to understand others' perspectives and empathize with them.

You're right. It's not so much the fact that their mothers work that causes kids today to suffer. Whether gathering food or earning a wage, women have always had to make a living. The problem is more that extended families have dissolved. When children can grow up with a grandmother, they grow faster, are more intelligent, and even have a better attachment to their mother. All that has been well established. And just think of President Obama. He grew up with his mother, his grandmother, and various other attachment figures.

He came from what would conventionally be called a broken home.

Yes, but also an extended family where he never doubted the commitment of his grandparents or his mother, while also having to adapt to a wide range of social experiences. I believe this enabled him to develop his extraordinary talent for integrating multiple perspectives. Apparently, he drew similar conclusions from that experience. I think the Obamas were very wise to have the maternal grandmother move into the White House with them and their children.

Unfortunately, the extended family model would hardly work for everyone. For many people, the mere thought of living under one roof with their mother-in-law is a nightmare.

Of course, we also need other solutions so that our children can again grow up with a greater number of attachment figures. But why shouldn't we make better use of the availability of older people? Nurturing children has always been a job for postreproductive humans. The aging of our society does not necessarily have to be a bad thing. It could conceivably pay off in more care. What seems far more nightmarish to me are worlds where resources are increasingly directed towards the old at the expense of the young.

Do you yearn to be a grandmother yourself?

Absolutely! But my children have told me not to hold my breath. They have other plans first. But I agree with them. Parenthood brings tremendous responsibilities. It's important to have your ducks in a row first.

Babies Can Be Smarter Than Us

Developmental psychologist
Alison Gopnik
on childhood

◆

Most people view their path from childhood to adulthood as progress, but for the American developmental psychologist Alison Gopnik it's also a story of loss. She compares children with butterflies that turn into caterpillars—our minds were once able to fly, whereas now they crawl along on the ground.

What happened to us? For three decades, Gopnik has investigated the way babies and children think. But advice for parents, teachers, or even her audience at the Davos World Economic Forum is, in her view, only a by-product of her research. Gopnik seeks less to understand childhood than human life as a whole. At the University of California, Berkeley, she is a professor of psychology and philosophy.

We meet in her old wooden house, a stone's throw from the campus. Ever since her own three sons moved out, it has been strangely quiet here, Gopnik says. A visit from her first grandson, a four-month-old, is planned for the afternoon.

◆

Professor Gopnik, would you like to be a child again?

I had a very happy childhood. And for a week I would like to see the world again through the eyes of a three-year-old. That would make my work easier. But permanently? I don't think I could stand the intense emotions. Imagine you're in Paris for the first time, you're in the midst of a tormented love affair, and you've just smoked a pack of Gauloises and had three espressos—that's what it's like to be a baby. Neuroscience suggests as much. That state strikes me as pretty exhausting.

Why do you call a three-year-old a baby?

For me, a baby is anyone who has chubby cheeks and funny pronunciation—so everyone under five. In English there's no single term for that stage of life.

In German we say *Kleinkind*, literally a small child.

That's an advantage of your language. On the other hand, "baby" captures our feelings for those little creatures really well—our special affection, our concern for them. That's why we also call adults we love with particular tenderness "baby."

My two-year-old son and my four-year-old daughter would be indignant about that! They disparagingly call any child even half a year younger "baby."

True. Many of the most popular children's stories are about their longing to be out in the world on their own without grown-ups.

Like Pippi Longstocking, or Mowgli in *The Jungle Book*.

They want to be independent, escape their childhood. Only once we're grown-ups can we afford the luxury of idealizing that time.

And yet we can't get back to it even in our imagination. You've written that "children and adults are different forms of *Homo sapiens*." What makes you think that?

Many people regard children as defective adults. That view is shared by teachers, neuroscientists, philosophers. Even Jean Piaget, the great pioneer of developmental psychology, who was the first to take children's minds seriously, described their defects more than anything else—not the advantages children have over us. It's far more likely, however, that in each stage of life nature accepts certain weaknesses as a trade-off for special strengths. That's where the differences come from. Caterpillars aren't defective butterflies.

Still, our children resemble us.

Externally. But why is developmental psychology so much fun? Because there are no Martians. The next best thing, if you want to investigate an alien intelligence, is these creatures with small bodies and big heads. Literally everything about a two-year-old is different from what we assume. And these aliens control us— often without our realizing it.

Surprisingly, you claim that children are more conscious than adults. What exactly do you mean by that?

According to philosophers, there are different kinds of consciousness—an awareness of our own internal state and an awareness of the external world. The first is what Descartes had in mind with his statement "I think, therefore I am." But if we abandon ourselves completely to our thoughts and feelings, we block out our surroundings. When I was absorbed in my work, my children would have fun shouting things like "Mommy, there's a puma in the yard!" Then they were delighted to get the usual response of "All right, honey" from me. I hadn't even heard what they were saying.

The proverbial absentminded professor.

And yet we regard that as the highest form of consciousness. Then there's the inverse of that, when we are completely spellbound by our environment and shut out the internal chatter and self-consciousness. That's what babies do. Now is that more consciousness or less? I think it's at least a state of greater awareness.

At the zoo, my four-year-old daughter spots every reptile, no matter how perfectly camouflaged. Her eight-year-old sister, on the other hand, who is something of a dreamer, sees only an empty terrarium. Maybe that's not only a function of the two girls' different personalities, but also of their ages.

The idea that little ones perceive more is supported not only by experiments—but also by what a store detective once told me. He looks out over his store from a balcony, where no adults notice him. But the children under five wave to him. Here you have an example of how much we can learn from children. Unfortunately, the typical philosopher still sits alone in his armchair and thinks about his consciousness, while the wealth of conscious states in other creatures eludes him.

But most people mainly want to find out about themselves. If children perceive, think, and feel in a completely different way than adults, can they really contribute to solving the mysteries of our consciousness?

Increasingly, biologists have recognized that we can understand an organism only if we understand the stages of its development. I think the same goes for our minds. How much of our knowledge of the world is innate to us? How much of it must be learned? Where does our sense of morality come from? Questions like these can be answered only if we understand our childhood.

The question of whether something like innate knowledge exists goes back a long time. Even the ancient Greek philosophers reflected on it.

At this point, it's clear that babies are true masters at making connections. Even one-year-olds do something like unconscious statistics: They can differentiate frequent from infrequent events and derive rules from that. And three-year-olds have a conception of cause and effect. They acquire that by playing around with anything they can get their hands on.

By that account, we wouldn't have innate knowledge—but we would probably have innate rules for how we structure our experiences. Whatever we encountered we would try to press into an unconscious schema of probability, cause, and effect. Anything that didn't fit into that framework would remain concealed from us. That's just what Immanuel Kant posited in the eighteenth century.

Yes, but the categories aren't rigid. A nine-month-old baby understands probability differently than an eighteen-month-old or an adult. Basically, children explore the world the way scientists do; their theories are constantly changing.

It's comforting to know that we're witnessing scientific research when the children once again make a mess of the house. But were you as the mother of three boys able to keep your cool from such a philosophical vantage point?

I'm absentminded and disorganized anyway. That made it easy to live with little kids at the same time as I was trying to advance my scientific career. I found the chaos all around me completely natural. That might be harder for Germans.

Well....

A stereotype, I know. But children's play is actually highly rational. We now know that a little bit of disorder often leads to better learning results than a systematic approach. The less you know about a problem, the better random trial and error works. Children and scientists discover things more quickly that way than with thought-out experiments. In the early years of life, consciousness resembles a lantern—it illuminates everything it encounters. Its only goal is to find out as much as possible about the world. Later, however, when we have to produce results, our awareness is directed like a spotlight.

The way children think reminds me of Leonardo da Vinci, who was always dealing with a dozen intricate problems at the same time, from painting to hydraulic engineering to the construction of flying machines—though he managed to practically implement only very few of his insights. Is there a Leonardo in every child?

Absolutely. We're always hearing how important it is to teach children focus. But impulse control comes at the cost of creativity. We know that from studies of jazz musicians. When they're improvising, their brains work in a completely different way than when they're playing from a score: The centers that focus attention are inhibited. Human beings have managed to discover so much only because their minds have passed through that long uncontrolled stage.

When do we lose our broad view of the world?

That begins around the age of five. It's no accident that children almost everywhere start school at that time.

In the classroom, children are trained to practice goal-directed thinking. It's often said of Leonardo that he was a genius, even though he went to school for barely four years and hadn't even learned fractions. But perhaps that very fact was his good fortune—he was able to maintain his childlike thinking!

I've given a few talks to high-ranking physicists at research centers. I told them that scientists are big children.

How did the physicists respond to that?

They agreed. But, of course, not everyone can be a Peter Pan. We also need people who think in a goal-oriented fashion.

Our culture certainly gives precedence to results. It seems to me that more random childlike thinking could be beneficial and even useful—not to replace goal-directedness, but to complement it.

That's true, only most adults really have to strain to achieve a lantern consciousness. Certain forms of meditation can help us, or traveling to new places where we make discoveries aimlessly—even sabbatical years. Children, on the other hand, are just naturally in that state.

If we don't drive it out of them.

That's why I worry about what will happen to early childhood education. The current pressure on preschools to offer academic instruction is dangerous.

It usually comes from parents who want their children to be able to read or to speak Mandarin with no accent before the first day of school.

A mother in New York recently sued a preschool. She thought her three-year-old was playing too much there and was not being prepared enough for college! And then we professors are surprised when we get students who work hard but think only about what's going to be on the test. Well, we selected for those very kids. It would be better for us to favor young people who tell us that they failed an important test because they spent the night before discussing the meaning of life until dawn—because that's the attitude at the heart of philosophy and science.

Ironically, many ambitious parents who want smart children can end up limiting their children's potential.

That's true. Children are amazingly sensitive to whether adults are doing something in order to teach them, and that sensitivity can actually narrow their range of exploration. In a series of experiments, my colleague Laura Schulz fiddled with a fairly complicated electronic toy in front of a group of four-year-olds. Left alone, the children soon found out all the different things the toy could do. Schulz then directly showed other boys and girls of the same age a few functions of the toy. But when she left, those children only repeated the little bit she had demonstrated to them.

> What children learn by playing can't be foreseen or controlled. Apparently, ambitious parents lack confidence in their kids' learning ability.

The problem goes even deeper. It used to be a fact of life that people practiced taking care of children on their siblings, cousins, nephews, and nieces. But the extended family is a thing of the past. So we're probably the first generation of parents who have our first more intensive experience with children when we ourselves have them. That generates enormous insecurity. On the other hand, goal-directed behavior is very familiar to us from school and work. Now we try to apply that model to family life . . .

> . . . and fail . . .

. . . because raising children is simply not a goal-oriented activity.

> Really? I certainly want my children to find their way to a fulfilling adult life—even if I don't know exactly what that means for them.

Exactly. Still, middle-class parents in particular agonize over questions like: Am I doing the right thing? What's going to be the outcome? What's he going to tell his shrink when he's thirty? Yet it's not so much what we teach our children that matters most. It's more about providing them a protected space in which they can explore on their own. So the other possibility is to experience what it's like to be a child—and to understand what the child needs right now. . . .

> Which developmental psychology unfortunately doesn't tell us—because you and your colleagues describe a theoretical average child of three or five, when in reality each child has his or her own personality.

We don't yet understand where the differences come from. My guess is that certain tendencies of children become self-reinforcing with time. In the beginning, there might be tiny variations, such as genetic differences—one child prefers to make larger movements, another somewhat smaller movements. The children then choose an environment for themselves that corresponds to that disposition.

So one of them would get better and better at sports, the other at drawing.

And the reactions of parents and teachers further reinforce those preferences. That might also explain many of the differences between boys and girls.

And yet, even though I've tried to get my daughters to take an interest in tools and technology, they haven't done so. My son, however, when he was only a year old, would get excited as soon as he saw me reach for a screwdriver.

Things like that are often especially frustrating for parents who want to break through gender roles. The idea that testosterone could draw men to tools is absurd, of course. But it's quite likely that the hormone could cause a propensity in boys for large motor activity. Dad would then simply be more interesting to them when he putters around with tools than when he's sitting at his desk. And by imitating that and becoming more and more skilled with tools himself, the little boy magnifies the initially small difference from his sisters.

According to that theory, it would be fundamentally impossible to separate the influence of genes from that of the environment.

Because people are constantly changing their environment, we can never know what effect particular genetic differences have. Another good example is attention deficit hyperactivity disorder, ADHD.

Hundreds of thousands of children are on Ritalin for that.

ADHD has genetic causes. But for our distant ancestors, it might not have made a difference. If anything, hyperactive people might have been better hunters. But if you put children with that genetic predisposition in school, you have a problem. And suddenly we're talking about a genetically determined disease. Yet the classroom environment in which ADHD first appears has existed for only about one hundred years.

Ritalin helps these children adapt to school.

Even more than that, it helps their parents—because when hyperactive children are on Ritalin, their parents clearly cope better with them. So far, however, there's no evidence that the medication has a more positive effect on success in school than, for example, behavioral therapy.

On the other hand, I can relate to parents who get a prescription because they're afraid a completely unfocused child will never be able to keep up in school.

But here in the United States, three-year-olds are given Ritalin—that's insane. However, there are certainly cases in which medication is necessary.

Especially since parents don't have an infinite capacity to support their children. Even under ordinary circumstances, children often need more than we can give them. For me, that's the painful side of being a father.

Parenting confronts us in everyday life with some of the deepest moral dilemmas we'll ever face. That's part of what makes our relationship with children so interesting from a philosophical perspective. In no other human relationship do we care even close to as much for another person. I love my husband and try to be a good wife to him. So I cook for him and listen to him. But if that were all I did for a baby, it would be child abuse! And children don't even notice our sacrifices. If they perceive the constant care as something special, that's even a sign that something is wrong.

That's where we get the feeling that we're never good enough as mothers and fathers, that we always owe our children more than we've been able to give them. Does parenthood demand the impossible from us?

People who did as much for strangers as all of us do for our children would be called saints. . . .

Due to their selflessness, not because they perform miracles.

I'm a Jewish atheist, but I believe living with a three-year-old is indeed a fast way to achieve a certain amount of saintliness.

I don't feel like much of a saint myself.

Even great saints rarely do. As for guilt—in America no one is supposed to feel guilty about anything. But I'm afraid it's a completely appropriate reaction to our enormous responsibility as parents.

Feminists would be less than thrilled with such ideas. For decades, women have been trying to learn to not always place themselves in the service of others—and now you come along with talk about appropriate guilt and saintliness!

Feminism has two sides: Along with the struggle against oppression, it's also always been about taking female experiences seriously. After all, women haven't been twiddling their thumbs for the last ten thousand years; they have been raising the entire population of the earth. And the insights they have attained from that are just as valuable as the tradition of mostly solitary male philosophers and theologians. That's the very reason I made my foray into developmental psychology: I wanted to help open up to philosophy a perspective to which it had been blind for too long. It suited me—as the oldest of six siblings I've spent only three years in my whole life without taking care of little kids.

What question so plagued you that you sought an answer in philosophy?

The question of where our knowledge of the world comes from. Given how little information we get through the senses, we know an incredible amount. For me, that remains the mystery of all mysteries. It goes back to Plato. I read him for the first time when I was ten years old.

Did you understand him?

My parents would never have said that we children couldn't understand a philosopher. They gave us any book, because they had the sensible view that we would find the parts that were accessible to us. By the way, a number of philosophers have told me that they began with Plato at that age or a little later.

He appeals to young readers because he writes so beautifully and vividly.

And between the ages of eight and ten, children typically begin to ask theological questions, like "Where did everything come from?" Studies show that they think about these things even when they grow up in an atheist environment.

Yes, even our children in Berlin ask about God—though your example doesn't strike me as particularly theological.

But the answers children come up with on their own are quite theological, such as "Someone must have created the universe." They follow a natural progression. Three-year-old babies already ask very fundamental questions—about, say, what's going on in someone else's mind, or why other people do what they do. So it's normal that over time children seek more and more comprehensive explanations—until eventually they wonder whether the whole world might have a purpose.

I know that I asked myself questions like that toward the end of my elementary school years. But I have no memory of what prompted them. Our own childhood seems to me like a dream after awakening: We can salvage some scenes from our memory, but the farther we go back, the murkier it becomes—as if the whole wealth of our early years were lost forever.

We don't even know why that is. Probably children under the age of four don't see themselves as a single entity moving through time. Lacking a conception of their past self and present self as part of the same story, they wouldn't have a solid basis for accumulating enduring memories.

Hasn't it ever made you sad to witness how quickly and irretrievably childhood goes by?

Yes, it has. In Japan there's the wonderful term *mono no aware*. It refers to the very particular beauty of the fleeting: the cherry blossom, for example, or the first snow. To enjoy it, we have to devote ourselves to it—and love passionately something we can't control or hold on to.

Love Is the Offspring of Knowledge

Artist

Leonardo da Vinci

on the beginning of modern scientific research

◆

Leonardo da Vinci wasn't the first scientist to arouse my fascination; my parents and my grandmother were. But his work has captivated me for longer than that of any other scientist. From the time I, at the age of eight, held in my hands a book on his inventions, I was under his spell. And it was Leonardo's drawings of whirlpools and the inside of the human body that led me to his art. Though other interests had repeatedly drawn my attention away from da Vinci, a chance visit to Milan—where I had been invited in the summer of 2006 to present one of my books—was enough to reawaken my enduring enthusiasm for the boldness, the inventiveness, and the perspicacity of this scientific pioneer. There, I visited the Leonardo da Vinci Museum of Science and Technology, and as I strolled through the halls, viewing the replicas of his machines, I was suddenly seized again by the desire I had felt as an eight-year-old: I wanted to know everything about this man. In the months that followed, I discovered to my surprise how much about Leonardo's scientific research still awaits thorough examination.

Two years later, my book *Leonardo's Legacy* came out in German (and in 2010 in English translation). While working on it, I felt as if I were in constant dialogue with da Vinci: I asked him questions, and his numerous surviving notebooks and manuscripts provided the answers. There were moments when I wished I had a time machine so that I could meet him face-to-face. Perhaps it would have gone something like this:

In the year 1514, Leonardo da Vinci is living in Rome. He has followed the call of Giuliano de' Medici, the powerful brother of the new Pope Leo X. The sixty-two-year-old master resides with the members of his workshop in the papal chambers overlooking the Belvedere courtyard of the Vatican. Having arrived from the future, I am led into a studio in a side wing, where several unfinished oil paintings are propped up on easels. One of them is a portrait of a young woman with a strange smile; her black eyes seem to gaze perpetually at the viewer. A few paces from the painting stands Leonardo, wrapped in a knee-length cloak. His hair is white and sparse, but his beard flows down to his chest. Leonardo's left hand trembles slightly, perhaps as a result of a stroke. He is working on a pyramid-shaped structure; inside it, hundreds of little mirrors are mounted.

....

◆

Hello, Master Leonardo.

Greetings.

What's that?

This is the burning mirror. And if you say that the mirror is cold and yet casts warm rays, I say in reply that the ray comes from the sun and thus, in passing through the mirror, must resemble its cause.[*]

You're harnessing solar energy.

Energy? I don't know that word. My mirror can concentrate so much power in a single point that water can be warmed in a heating tank, like those used in dye factories, or for a swimming pool.

Has it occurred to you that your concave mirror can also be used to view distant objects?

---

[*] Leonardo's answers are drawn from his manuscripts and notebooks.

Of course. Here is a cover. Open it so that only the light of a single planet falls through. The reflected image will then show you the nature of that planet.

Very impressive. You've probably never heard the term "telescope" either. . . .

No.

But with all due respect: Shouldn't you be spending your time painting? That portrait of the young woman there has accompanied you for years.

Yes, yes, Lisa del Giocondo. Her husband called her Mona Lisa. He commissioned the painting from me in 1506. Back in Florence. . . .

You left Florence eight years ago. The portrait still isn't done—while your rivals Michelangelo and Raphael have been acquiring one commission after another. Your new employer, the pope, even makes fun of your scientific interests in public. "Instead of painting," he proclaims, "that Leonardo would rather whip up varnish from oils and plants like an alchemist."

There are painters who work only with practical experience and judgment of the eye, but without the use of reason. They are like mirrors that reproduce all the objects set in front of them without knowledge of those objects. But whoever takes up practice without science is like a sailor who boards a ship without a compass and can never be certain where he is going. Without science, nothing can be done well in painting.

But many critics refuse to take you seriously as a scientist. They point out that you never even had any higher education.

I know well that some vain people will think they have reason to deride me as a man without learning. Foolish people! They go around pompously and self-importantly, adorned not with their own labors, but with those of others, and will not permit me my own. Don't they know that my theories are drawn from experience rather than the words of others? All our knowledge originates in our perceptions.

You regard your critics as dusty scholars who only regurgitate books, while you yourself strive to understand the world with your own eyes and with experiments?

Exactly. What trust can we place in the ancient philosophers, who tried to define the nature of life and the soul—while things that at any time could have been discovered and proven by experience remained unknown or misunderstood for many centuries? Please excuse me for a moment. [A handsome young man enters and introduces himself as Francesco Melzi, Leonardo's assistant. He takes his master aside and whispers something to him. Leonardo winces.]

What happened?

He has slandered me before the pope and at the hospital. He is hindering my anatomical work.

Who is "he?"

Johannes, that German who makes mirrors for me. He has been snooping around in my workshop for a long time and tells the whole world everything. Because of him I can do nothing in secret.

You tried hard to prevent word from spreading about your work on corpses. You dissected them at night.

Yes. But to attain a true and complete knowledge of the blood vessels, I had to dissect more than ten human bodies, destroying all the other parts, and removing all the flesh surrounding those vessels. And as a single body didn't last long enough, I had to proceed with several bodies in succession, until I had attained a complete knowledge.

Didn't that ever disgust you?

And how! Even if you have the necessary passion for anatomical investigations, you will perhaps be deterred by your stomach, and if that doesn't stand in your way, then you will perhaps be hindered by the fear of spending the night in the company of those quartered and flayed and horrible-looking corpses.

Why did you do that to yourself? A better understanding of the network of blood vessels doesn't do anything to help your painting.

I'm drawn by an irrepressible desire to see the vast abundance of strange forms created by inventive nature. In the midst of scientific research, I feel as if I have come to the mouth of a huge unknown cavern. I stand before the entrance to the earth's interior, look this way and that in order to discern what awaits me. But this is rendered impossible by the deep darkness within the cavern. And after a while, two emotions suddenly stir in me — fear and desire, fear of the dark, threatening cavern and desire to find out whether there might be something marvelous within.

Your curiosity has almost always triumphed—not only over your own fear, but also over society's taboos. You even flout the Christian prohibition against disturbing the peace of the dead. What do you say to the criticism that you gained those insights only by desecrating corpses?

Noble people naturally have the desire for knowledge. And when you contemplate this machine of ours, do not be distressed that you owe your knowledge to another's death — but rather praise our Creator for endowing our intellect with such excellent powers of perception.

You regard the human body as a machine?

Certainly. Arms and legs, even the bite force of individual teeth, follow the lever principle. Swollen muscles function like wedges. And tendons hold the thighs in the joints as shrouds hold up the mast of a ship.

You knew some of the people whose bodies you later dissected.

That was back in Florence at the hospital of Santa Maria Nuova. There an old man told me a few hours before his death that he had lived a hundred years and felt no bodily ailment except for weakness. And as he sat on his bed in the hospital, without any movement he departed this life. I then began an anatomical investigation to figure out the cause of such a peaceful death.

What did you discover?

The decline came from the failure of the blood in the artery that feeds the heart and the other parts of the body, which I found to be very dried-up, shrunken, and withered. Old people who are in good health die from lack of sustenance. The passage through the blood vessels becomes more and more constricted until the capillaries close completely.

We call that arteriosclerosis. In the time I come from, almost everyone lives to be as old as you, Master. There's no more plague, fewer wars, and almost all women survive childbirth . . . most people die in their old age as a result of arteriosclerosis. The phenomenon you discovered was first named in the eighteenth century, incidentally, by an Alsatian surgeon named Jean Lobstein.

Arteriosclerosis—I like the word. That's really how it is: The outer layer of human veins acts the same way as oranges—the older they get, the thicker the peel becomes. When I performed an autopsy on a two-year-old boy, I found everything to be completely different from the case of the old man.

If you'd left it at that . . . you'd probably have gotten into less trouble. But you had to study unborn children. Recently you completed a whole series of drawings that show the fetus growing in the womb. And alongside them we read remarks on the soul. What were you trying to say with that?

The unborn child is vivified and nourished by the life of the mother. And one and the same soul governs the two bodies, for this creature shares its desires and fears and pains with the mother.

You mean that two creatures whose bodies are linked must also share a soul, because soul and body are inseparably connected?

Yes. Every part is inclined to unite with its whole in order to escape its own deficiencies. And the soul desires to dwell in the body, because without its instruments it can neither think nor feel.

Master, that is heresy. According to the Church, the soul is bestowed on a human being at the moment of conception. And although the

soul is linked to the body, it is not dependent on it, because the soul is immortal. Your current Pope Leo X issued a bull sharply condemning the "despicable heretics" who dared to doubt Church doctrine on the immortality of the soul. Aren't you afraid?

I know my limits. Have I ever claimed to know exact details about the soul? Whatever it might be, it is a divine thing. I leave the definitions of the soul to the minds of the monks, the fathers of the people, who through inspiration know all secrets. Let Holy Scripture stand, for it is the highest truth.

You haven't always sounded so tame, Leonardo. Once you even disputed the biblical story of Genesis.

When I was still working at the court of Milan, peasants from the mountains of Parma and Piacenza came to my workshop. They had with them a large sack of shells and corals full of holes, which they had collected in their fields. Many of them were still in their original condition.

Knowing that you might be interested in such things, the peasants sold them to you. What did you conclude from this trove?

The shells came from a height of more than a thousand cubits. But the summits in that area are much higher. If you claim that the Flood rose seven cubits above the highest mountain . . .

The Bible says 15 cubits. . . .

. . . then those shells, which always reside near the seashore, would have to be found on those mountains. But they are found at the mountains' bases—indeed, all at the same level, layer upon layer. I have also found the petrified shells in the valley of the Arno near my birthplace of Vinci, in Lombardy, and in the mountains of Verona.

You wandered through Italy in search of fossils?

I've collected them everywhere. And wherever I went, I found similar layers. Because of these layers of shells, I believe that all Italy once gradually emerged from a great sea. The bed of the sea became mountain ridges. The story of the Flood simply cannot be true. And the scores of ignorant people who refuse to accept that only prove their stupidity and simplemindedness.

Leonardo, the year is 1514. Everyone takes the Bible literally. You're very hard on your contemporaries.

Some of my contemporaries ought to call themselves nothing more than a passage for food, producers of excrement, and fillers of cesspits, because nothing good is accomplished by them and because they leave behind nothing but full cesspits.

What makes you so angry?

Falsehood is so vile that even when it praises God's works, it is an affront to his divine grace. Truth is so excellent that it ennobles the lowest things merely by speaking of them. But those who do not strive for complete knowledge trespass against knowledge and love—for love of anything is the offspring of knowledge; the more certain the knowledge, the more fervent the love.

Ludovico Sforza, the Duke of Milan, didn't pay you for fifteen years for your love of truth. He wanted a gigantic equestrian statue from you.

Most of all he wanted weapons.

And you were to construct them.

In my application letter, I promised the duke catapults of wonderful efficacy and cannons that could produce a hail of stones, instilling great terror in the enemy with their smoke and causing much loss and confusion. And unassailable covered wagons, which would enter the enemy's ranks with their artillery and rout even the greatest crowd of armed men.

We call them tanks. But you worked on far more terrible methods of warfare. You planned a poison war.

Yes. A powder of chalk, arsenic, and verdigris could be thrown with small catapults at enemy ships; all those who inhaled the powder would die. I warned of the need to ensure that no wind blows the powder toward you, or that you at least cover your nose and mouth with a fine wet cloth so that the powder cannot enter.

Is it really true that you refused to touch meat lest you harm any creature?

Well, doesn't nature produce enough simple foods for you to satisfy yourself?

And you've called war "bestial madness?"

It is! If you have seen the marvelous works of nature and you judge it to be an atrocious act to destroy them, then consider how infinitely atrocious it is to take a human life.

So how, then, could you design weapons of mass destruction?

To preserve the chief gift of nature, which is freedom, I invented means of attack and defense in case we should be besieged by ambitious tyrants.

You can't be serious. Ludovico Sforza, your Milanese patron, was one of the most expansionist tyrants of his time. And when he fell, you entered the service of Cesare Borgia—the son of Pope Alexander VI, who terrorized half of Italy with his troops. You became his chief engineer. For almost a year, you traveled with this man, who was feared by all for his extraordinary brutality.

Valentino. . . .

That's what you called him.

Where is he now?

After his father, the pope, died and he was exiled, he was killed in an ambush in Spain.

He had promised me an Archimedes manuscript from the library of the bishop of Padua. He still owes me my pay as well.

When you returned after your expeditions with Borgia to your hometown of Florence, you literally had to start from scratch—at over fifty years old. You were widely praised as an artist, but you no longer got along with the clients in this city-state. It must have been a difficult time. You were living off money you had saved.

A severed tree that would sprout again: I was still hopeful.

You slammed the door in worthy customers' faces. You even rejected the art patron Isabella d'Este, Marchesa of Mantua—even though she practically begged you for a painting. She was informed only that the Master was too busy with mathematical and scientific studies to paint. You were devoting yourself to attempts to square the circle—and to fly.

For that I first had to understand bird flight—because birds are flying machines.

For days you roamed the hills overlooking Florence and observed birds of prey.

To describe clearly the kite that lives there seemed to be my destiny, for I recall as my earliest childhood memory that while I was in my cradle, a red kite came and opened my mouth with its tail.

That story is well known. Sigmund Freud, a psychoanalyst highly regarded by many of my contemporaries and, of course, unknown to you, wrote a whole book about you in 1910. Unfortunately, he mistranslated the word for "kite" from your notebooks as "vulture." Freud interpreted your memory as a coded homosexual fantasy.

Nonsense. I simply wanted to fly. Two decades earlier, in Milan, I had already planned the first flying machines. I even made notes on how I could test my designs at Corte Vecchia, the ducal palace, without drawing too much attention. My plan was to board up the upper room and go up to the roof. And I would remain on the side of the tower so that the builders working on the unfinished cupola of the nearby cathedral wouldn't see me.

How did your Milan attempts work out?

No comment.

When you returned to Florence after your adventures with Borgia, you apparently hoped to take off from Monte Ceceri with a flying device. The name of the hill overlooking the city translates as "Swan Mountain."

I hoped that the great bird would take its first flight upon the back of the great swan, filling the whole world with amazement and all writings with its fame and bringing eternal glory to the place where it was born.

Monte Ceceri rises more than one thousand feet over Florence. Didn't you find it dangerous to take off from there?

Wineskins, inflated with air and tied together like the beads of a rosary, are wrapped around the aviator by a helper. With these, a man falling from a height of six cubits will not do himself any harm, whether he falls into water or on land. But if you should fall with the double chain of wineskins tied underneath you, make sure that these are what first strike the ground.

Your pneumatic protective suit must have served you well. The physician and natural philosopher Gerolamo Cardano wrote several years later about your attempts: "Leonardo tried to fly and failed." Do you know how feasible some of your designs actually turned out to be? Shortly before I departed my century to visit you, people flew with slightly altered replicas of them.

Really?

I saw the wings myself. English aircraft engineers constructed them according to the plans you left behind.

How far did they fly?

More than six hundred feet. That's about four hundred cubits. Do you know that we people of the twenty-first century are just beginning to read substantial portions of your notes correctly? One reason we were groping in the dark for so long is that, until recently, it was mainly our art historians who dealt with your work. There wasn't much they could do with the hydraulic experiments, the laws of aerodynamics, or the investigations of anatomy.

Let no one who is ignorant of the mathematical sciences read my works.

Only in the years before my departure to visit you did scientists in the fields you explored begin to study your writings systematically.

They brought to light the most amazing things. For example, you invented artificial heart valves and drew accurately the blood flow in the human heart. . . .

Yes, I constructed a heart out of glass, filled it with water, threw in millet seeds, and observed them.

Scientists of my era didn't manage to investigate that blood flow until 1998. They used computed tomography. Other experts even assert that you designed robots and the first digital computer. Sorry I have to use words you don't know from the very distant future. Are you amazed that we have understood some of your ideas only five hundred years after your death?

It doesn't surprise me much. One of my life's maxims was: Avoid those works of which the results die with the worker—for we do not lack ways of measuring our meager days; therefore it should be our pleasure not to squander them vainly or spend them ingloriously, leaving behind no memory in the minds of mortals.

Your notebooks and writings are full of exercises you recommend—and probably used yourself—for the training of the human mind. Can anyone enhance the capacity of their intellect?

I'll tell you a little story. One day a razor blade saw the sun glittering on its surface. Full of pride, it decided not to return to work at the barbershop and looked for a peaceful hiding place. After a few months, when it emerged again into the light of day, it noticed that all its radiance had disappeared, because a layer of rust covered it. The same thing happens with the human mind: We must use it constantly. As soon as we abandon ourselves to idleness, the mind, like the razor blade, loses its sharp edge, and an ugly rust of ignorance disfigures it. Farewell!

Acknowledgments

Declining a visit from admirers of his work, the Hungarian composer György Ligeti is supposed to have said, "To enjoy a goose liver pâté, you don't have to know the goose." It's true that Ligeti was not only one of the twentieth century's most significant creators of music, but also a brilliant mind in many other respects—about this, however, he was wrong. In many cases, I had followed the careers of my conversation partners for years before we first met, and I knew all their most important works. And yet every single encounter was enriching for me in a way that even the most thorough reading could not have been. For one thing, meeting people helps you better understand their ideas. There was also the joy of moments when the conversation led to insights that were new to both of us. And finally, I shared with the scientists memorable personal experiences, including, among many others, exploring with Vittorio Gallese the cuisine of his native city of Parma and watching at Jared Diamond's house the historic television broadcast of the inauguration of Barack Obama. I am profoundly grateful to my conversation partners for the openness and engagement with which they participated in these unusually long and intense dialogues.

This project would not have been possible without the continuous support of my editors at the *Zeit Magazin*, in particular Christoph Amend, Florian Illies, Stephan Lebert, and Matthias Stolz. Many thanks to my agents, Matthias Landwehr, who made every effort to ensure that the conversation series in the magazine and later the book would become a reality, and Frank Jakobs, who handled foreign rights. With my German publisher, S. Fischer Verlag, I felt, as always, in the best hands. I'm indebted to Nina Bschorr and Peter Sillem for their usual great editorial advice on the German version. This is the fourth book I've worked on with my American publisher, Matthew Lore, and once again the experience has been extremely gratifying. My gratitude goes to Ross Benjamin for the outstanding English translation and to Nicholas Cizek for the sharp-sighted editing.

Finally, I thank my beloved wife, Alexandra Rigos, for her help and support—and not least of all for her flair for distinguishing good ideas from better ones.

Photograph Credits

MARTIN REES: Photo by Dominik Gigler

RICHARD DAWKINS: Photo © Immo Klink

V. S. RAMACHANDRAN: Photo by Beatrice Ring

JARED DIAMOND: Photo by Gabor Ekecs

JANE GOODALL: Photo by Markus Burke

STEVEN WEINBERG: Photo by Steven Noreyko

ELIZABETH BLACKBURN: Photo by Daniel Schumann

PETER SINGER: Photo by Denise Applewhite/Princeton University

NICHOLAS CHRISTAKIS: Photo by Giorgos Moutafis

CRAIG VENTER: Photo by the J. Craig Venter Institute

ROALD HOFFMANN: Photo by Clemens Loew

HANNAH MONYER: Photo by Claus Geiss

VITTORIO GALLESE: Photo by Albrecht Tubke

RAGHAVENDRA GADAGKAR: Photo by Patrick Voigt

ERNST FEHR: Photo by Lena Amuat

WALTER ZIEGLGÄNSBERGER: Photo by Andy Rumball

SARAH HRDY: Photo by Regine Petersen

ALISON GOPNIK: Photo by Daniel Schumann

LEONARDO DA VINCI: Image courtesy the Biblioteca Reale, Turin

About the Author

Stefan Klein, PhD, has studied physics and analytical philosophy and holds a doctorate in biophysics. After several years as an academic researcher, he turned to writing about science for a general audience. From 1996 to 1999 he was an editor at *Der Spiegel*, Germany's leading news magazine, and in 1998 he won the prestigious Georg von Holtzbrinck Prize for Science Journalism. Today Klein is recognized as one of Europe's most influential science writers and journalists. His interviews with the world's leading scientists are a regular feature in Germany's *Zeit Magazin*. His books, which have been translated into more than twenty-five languages, include *Survival of the Nicest*, the #1 international bestseller *The Science of Happiness*, *The Secret Pulse of Time*, and *Leonardo's Legacy*. A frequent speaker and university guest lecturer, he lives with his family in Berlin.

Translator Ross Benjamin is a 2015 Guggenheim Fellow. He has received the Helen and Kurt Wolff Translator's Prize as well as a National Endowment for the Arts Literature Fellowship for Translation. He is currently at work on a translation of Franz Kafka's complete *Diaries*, to be published by Liveright/Norton.